HOW to LIE
with MAPS

HOW to LIE with MAPS

Second Edition

MARK MONMONIER

The University of Chicago Press
Chicago and London

The University of Chicago Press, Chicago 60637
The University of Chicago Press, Ltd., London
© 1991, 1996 by The University of Chicago
All rights reserved. Second edition published 1996
Printed in the United States of America

12 11 10 09 08 07 06 05   4 5

ISBN: 0-226-53421-9 (paper)

Library of Congress Cataloging-in-Publication Data

Monmonier, Mark S.
    How to lie with maps / Mark Monmonier. — 2nd ed.
    p.   cm.
    Includes bibliographical references and index.
       1. Cartography. 2. Deception. I. Title.
G108.7.M66   1996   95–32199
526—dc20   CIP

For Marge and Jo

CONTENTS

5

MAPS THAT ADVERTISE
58

6

DEVELOPMENT MAPS
(OR, HOW TO SEDUCE THE TOWN BOARD)
71

7

MAPS FOR POLITICAL PROPAGANDA
87

8

MAPS, DEFENSE, AND DISINFORMATION:
FOOL THINE ENEMY
113

9

LARGE-SCALE MAPPING, CULTURE, AND THE NATIONAL INTEREST

123

10

DATA MAPS: MAKING NONSENSE OF THE CENSUS

139

11

COLOR: ATTRACTION AND DISTRACTION

163

12

MULTIMEDIA, EXPERIENTIAL MAPS, AND GRAPHIC SCRIPTS

174

13

EPILOGUE

184

FOREWORD

As a nation, Americans may not be as geographically literate as they should be, but they are fascinated by maps. I have the evidence. Every time I use maps to make a point during one of my geography segments on "Good Morning America," viewers want copies of the maps. They write that the maps provided a new perspective on an issue they had not thought about that way. They complain that the maps were not on screen long enough. And sometimes a sharp-eyed viewer will detect an error. Did I draw Krajina smaller than it really is, to make it look weaker in the wars of former Yugoslavia? Why were the Golan Heights and the West Bank given the same color when their status is different? Was that maritime boundary between Japan and South Korea in the right place?

This is the book to which I refer such map enthusiasts. Mark Monmonier has written a delightful, provocative, informative book for people who enjoy looking at and using maps, and especially for those who, like these viewers, "read" maps attentively. If a picture is worth a thousand words, a map can be worth a million—but beware. All maps distort reality. All mapmakers use generalization and symbolization to highlight critical information and to suppress detail of lower priority. All cartography seeks to portray the complex, three-dimensional world on a flat sheet of paper or on a television or video screen. In short, the author warns, all maps must tell white lies.

And sometimes these lies are not so little. Maps are informative, but they also can be deceptive, even threatening. It probably is safe to say that all of us have been misled, at one time or another, by a map designed to hide something the mapmaker did not want us to know, or drawn in such a way that we jump to false conclusions from it. Whether you are buying a house, positioning a business, selecting a school, or planning a vacation, maps form an essential part of the process. Some of those maps, however, may pull the wool over your eyes. They cross the line between information and advocacy. *How to Lie with Maps* constitutes a primer on such misuse, prepares the reader to deal with it, and entertainingly guides you through the cartographic jungle.

In our world of changing political and strategic relationships and devolving nation-states, maps become propaganda tools. Turkish Cypriots, Sri Lankan Tamils, Crimean Russians publish maps that proclaim their political aspirations, fueling nationalisms that spell disaster for the state system. Some national governments even go so far as to commit cartographic aggression, mapping parts of neighboring countries as their own. Well before Iraq invaded Kuwait, official Iraqi maps had shown Kuwait as Baghdad's nineteenth province. Chinese maps today incorporate parts of what, on standard world maps, is northern India. Argentina, reports the author, prints postage stamps showing hegemony over a sector of Antarctica that includes Chilean as well as British claims.

This, therefore, is a significant book, at the international as well as the individual level. Dr. Monmonier attaches great importance to the public understanding of maps as a medium of communication, equating it with the sort of knowledge that allows us to control fire and use electricity. And, indeed, *How to Lie with Maps* is powerful evidence of the cost of ignorance— that lingering "cartographic mystique" that is part of our continuing geographic illiteracy. Greater awareness of the kind the author advocates in this book might have saved the United States from the debacle of Indochina. More currently, a wider understanding of what temperature maps can and cannot reveal might temper the rush to global "greenhouse" scenarios.

This fascinating volume deals with such serious issues in a lively, often humorous, always engrossing way. Read it, and the maps you view henceforth will have new meaning.

H. J. de Blij

ACKNOWLEDGMENTS

Many friends, colleagues, and authors have helped hone my interest in maps into the wary enthusiasm I hope to communicate in this book. I owe a special debt of gratitude

to Darrell Huff, who demonstrated with *How to Lie with Statistics* that a cogent, well-illustrated essay on a seemingly arcane topic could be both insightfully informative and enjoyably readable;

to Syracuse University, for a one-semester sabbatical leave during which much of the book was written;

to Mike Kirchoff, Marcia Harrington, and Ren Vasiliev, for supplementary artwork and for critical comments on the design and execution of my illustrations;

to John Snyder, for information and insights about map projections;

to David Woodward, for continued encouragement; and

to Penny Kaiserlian, for the opportunity to expand and add color.

Chapter 1

INTRODUCTION

Not only is it easy to lie with maps, it's essential. To portray meaningful relationships for a complex, three-dimensional world on a flat sheet of paper or a video screen, a map must distort reality. As a scale model, the map must use symbols that almost always are proportionally much bigger or thicker than the features they represent. To avoid hiding critical information in a fog of detail, the map must offer a selective, incomplete view of reality. There's no escape from the cartographic paradox: to present a useful and truthful picture, an accurate map must tell white lies.

Because most map users willingly tolerate white lies on maps, it's not difficult for maps also to tell more serious lies. Map users generally are a trusting lot: they understand the need to distort geometry and suppress features, and they believe the cartographer really does know where to draw the line, figuratively as well as literally. As with many things beyond their full understanding, they readily entrust mapmaking to a priesthood of technically competent designers and drafters working for government agencies and commercial firms. Yet cartographers are not licensed, and many mapmakers competent in commercial art or the use of computer workstations have never studied cartography. Map users seldom, if ever, question these authorities, and they often fail to appreciate the map's power as a tool of deliberate falsification or subtle propaganda.

Because of personal computers and electronic publishing, map users can now easily lie to themselves—and be unaware of it. Before the personal computer, folk cartography consisted largely of hand-drawn maps giving directions. The direction giver had full control over pencil and paper and usually had no difficulty transferring routes, landmarks, and other relevant recollections from mind to map. The computer allows programmers, marketing experts, and other anonymous middlemen without cartographic savvy to strongly influence the look of the map and gives modern-day folk maps the crisp type, uniform symbols, and verisimilitude of maps from the cartographic priesthood. Yet software developers commonly have made it easy for the lay cartographer to select an inappropriate projection or a misleading set of symbols. Because of advances in low-cost computer graphics, inadvertent yet serious cartographic lies can appear respectable and accurate.

The potential for cartographic mischief extends well beyond the deliberate suppression used by some cartographer-politicians and the electronic blunders made by the cartographically ignorant. If any single caveat can alert map users to their unhealthy but widespread naïveté, it is that *a single map is but one of an indefinitely large number of maps that might be produced for the same situation or from the same data.* The italics reflect an academic lifetime of browbeating undergraduates with this obvious but readily ignored warning. How easy it is to forget, and how revealing to recall, that map authors can experiment freely with features, measurements, area of coverage, and symbols and can pick the map that best presents their case or supports their unconscious bias. Map users must be aware that cartographic license is enormously broad.

The purpose of this book is to promote a healthy skepticism about maps, not to foster either cynicism or deliberate dishonesty. In showing how to lie with maps, I want to make readers aware that maps, like speeches and paintings, are authored collections of information and also are subject to distortions arising from ignorance, greed, ideological blindness, or malice.

Examining the misuses of maps also provides an interesting introduction to the nature of maps and their range of appropriate uses. Chapter 2 considers as potential sources of distortion the map's main elements: scale, projection, and symbolization. Chapter 3 further pursues the effects of scale by examining the various white lies cartographers justify as necessary generalization, and chapter 4 looks at common blunders resulting from the mapmaker's ignorance or oversight. Chapter 5 treats the seductive use of symbols in advertising maps, and chapter 6 explores exaggeration and suppression in maps prepared for development plans and environmental impact statements. Chapters 7 and 8 examine distorted maps used by governments as political propaganda and as "disinformation" for military opponents. Government mapping is also central to Chapter 9, which investigates the effects of national culture and bureaucratic inertia on detailed topographic maps. The next two chapters are particularly relevant to users of mapping software and electronic publishing: chapter 10 addresses distortion and self-deception in statistical maps made from census data and other quantitative information, and chapter 11 looks at how a careless or Machiavellian choice of colors can confuse or mislead the map viewer. Chapter 12 peers ahead toward a future in which dynamic, highly customized maps promote exploration and interpretation. Chapter 13 concludes by noting maps' dual and sometimes conflicting roles and by recommending a skeptical assessment of the map author's motives.

A book about how to lie with maps can be more useful than a book about how to lie with words. After all, everyone is familiar with verbal lies, nefarious as well as white, and is wary about how words can be manipulated. Our schools teach their pupils to be cautious consumers who read the fine print and between the lines, and the public has a guarded respect for advertising, law, marketing, politics, public relations, writing, and other occupations requiring skill in verbal manipulation. Yet education in the use of maps and diagrams is spotty and limited, and many otherwise educated people are graphically and cartographically illiterate. Maps, like numbers, are often arcane images accorded undue respect and credibility. This book's principal goal is to dispel this cartographic mystique and promote a more informed use of maps based upon an understanding and appreciation of their flexibility as a medium of communication.

The book's insights can be especially useful for those who might more effectively use maps in their work or as citizens fighting environmental deterioration or social ills. The informed skeptic becomes a perceptive map author, better able to describe locational characters and explain geographic relationships as well as better equipped to recognize and counter the self-serving arguments of biased or dishonest mapmakers.

Where a deep mistrust of maps reflects either ignorance of how maps work or a bad personal experience with maps, this book can help overcome an unhealthy skepticism called *cartophobia*. Maps need be no more threatening or less reliable than words, and rejecting or avoiding or ignoring maps is akin to the mindless fears of illiterates who regard books as evil or dangerous. This book's revelations about how maps *must* be white lies but may *sometimes* become real lies should provide the same sort of reassuring knowledge that allows humans to control and exploit fire and electricity.

Chapter 2

ELEMENTS OF THE MAP

Maps have three basic attributes: scale, projection, and symbolization. Each element is a source of distortion. As a group, they describe the essence of the map's possibilities and limitations. No one can use maps or make maps safely and effectively without understanding map scales, map projections, and map symbols.

Scale

Most maps are smaller than the reality they represent, and map scales tell us how much smaller. Maps can state their scale in three ways: as a ratio, as a short sentence, and as a simple graph. Figure 2.1 shows some typical statements of map scale.

Ratio scales relate one unit of distance on the map to a specific distance on the ground. The units must be the same, so that a ratio of 1:10,000 means that a 1-inch line on the map represents a 10,000-inch stretch of road—or that 1 centimeter represents 10,000 centimeters or 1 foot stands for 10,000 feet. As long as they are the same, the units don't matter and need not be stated; the ratio scale is a dimensionless number. By convention, the part of the ratio to the left of the colon is always 1.

Some maps state the ratio scale as a fraction, but both forms have the same meaning. Whether the mapmaker uses 1:24,000 or 1/24,000 is solely a matter of style.

Fractional statements help the user compare map scales. A scale of 1/10,000 (or 1:10,000) is larger than a scale of 1/250,000 (or 1:250,000) because 1/10,000 is a larger fraction than 1/250,000. Recall that small fractions have big denomi-

Ratio Scales	Verbal Scales
1:9,600	One inch represents 800 feet.
1:24,000	One inch represents 2,000 feet.
1:50,000	One centimeter represents 500 meters.
1:250,000	One inch represents (approximately) 4 miles.
1:2,000,000	One inch represents (approximately) 32 miles, one centimeter represents 20 kilometers.

Graphic Scales

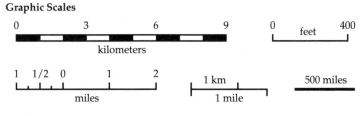

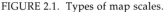

FIGURE 2.1. Types of map scales.

nators and big fractions have small denominators, or that half (1/2) a pie is more than a quarter (1/4) of the pie. In general, "large-scale" maps have scales of 1:24,000 or larger, whereas "small-scale" maps have scales of 1:500,000 or smaller. But these distinctions are relative: in a city planning office where the largest map scale is 1:50,000, "small-scale" might refer to maps at 1:24,000 or smaller and "large-scale" to maps at 1:4,800 or larger.

Large-scale maps tend to be more detailed than small-scale maps. Consider two maps, one at 1:10,000 and the other at 1:10,000,000. A 1-inch line at 1:10,000 represents 10,000 inches, which is 833 1/3 feet, or roughly 0.16 miles. At this scale a square measuring 1 inch on each side represents an area of .025 mi², or roughly 16 acres. In contrast, at 1:10,000,000 the 1-inch line on the map represents almost 158 miles, and the square inch would represent an area slightly over 24,900 mi², or nearly 16 million acres. In this example the square inch on the large-scale map could show features on the ground in far greater detail than the square inch on the small-scale map. Both maps would have to suppress some details, but the designer of the 1:10,000,000-scale map must be far more selective than the cartographer producing the 1:10,000-scale map. In the sense that all maps tell white lies about the planet, small-scale maps have a smaller capacity for truth than large-scale maps.

Verbal statements such as "one inch represents one mile" relate units convenient for measuring distances on the map to units commonly used for estimating and thinking about distances on the ground. For most users this simple sentence is more meaningful than the corresponding ratio scale of 1:63,360, or its close approximation, 1:62,500. British map users commonly identify various map series with adjective phrases such as "inch to the mile" or "four miles to the inch" (a close approximation for 1:250,000).

Sometimes a mapmaker might say "equals" instead of "represents." Although technically absurd, "equals" in these cases might more kindly be considered a shorthand for "is the equivalent of." Yet the skeptic rightly warns of cartographic seduction, for "one inch equals one mile" not only robs the user of a subtle reminder that the map is merely a symbolic model but also falsely suggests that the mapped image is reality. As later chapters show, this delusion can be dangerous.

Metric units make verbal scales less necessary. Persons familiar with centimeters and kilometers have little need for sentences to tell them that at 1:100,000, one centimeter represents one kilometer, or that at 1:25,000 four centimeters represent one kilometer. In Europe, where metric units are standard, round-number map scales of 1:10,000, 1:25,000, 1:50,000, and 1:100,000 are common. In the United States, where the metric system's most prominent inroads have been in the liquor and drug businesses, large-scale maps typically represent reality at scales of 1:9,600 ("one inch represents 800 feet"), 1:24,000 ("one inch represents 2,000 feet"), and 1:62,500 ("one inch represents [slightly less than] one mile").

Graphic scales are not only the most helpful means of communicating map scale but also the safest. An alternative to blind trust in the user's sense of distance and skill in mental arithmetic, the simple bar scale typically portrays a series of conveniently rounded distances appropriate to the map's function and the area covered. Graphic scales are particularly safe when a newspaper or magazine publisher might reduce or enlarge the map without consulting the mapmaker. For example, a five-inch-wide map labeled "1:50,000" would have a scale less than 1:80,000 if reduced to fit a newspaper column three inches wide, whereas a scale bar representing a half-mile would shrink along with the map's other symbols and distances. Ratio and verbal scales are useless on video maps, since television screens and thus the map scales vary widely and unpredictably.

Map Projections

Map projections, which transform the curved, three-dimensional surface of the planet into a flat, two-dimensional plane, can greatly distort map scale. Although the globe can be a true scale model of the earth, with a constant scale at all points and in all directions, the flat map stretches some distances and shortens others, so that scale varies from point to point. Moreover, scale at a point tends to vary with direction as well.

The world map projection in figure 2.2 illustrates the often severe scale differences found on maps portraying large areas. In this instance map scale is constant along the equator and the meridians, shown as straight lines perpendicular to the equator and running from the North Pole to the South Pole. (If the terms *parallel, meridian, latitude,* and *longitude* seem puzzling, the quick review of basic world geography found in the Appendix might be helpful.) Because the meridians have the same scale as the equator, each meridian (if we assume the earth is a *perfect* sphere) is half the length of the equator. Because scale is constant along the meridians, the map preserves the even spacing of parallels separated by 30° of latitude. But on this map all parallels are the same length, even though on the earth or a globe parallels decrease in length from the equator to the poles. Moreover, the map projection has stretched the poles from points with no length to lines as long as the equator. North-south scale is constant, but east-west scale increases to twice the north-south scale at 60° N and 60° S, and to infinity at the poles.

Ratio scales commonly describe a world map's capacity for detail. But the scale is strictly valid for just a few lines on the map—in the case of figure 2.2, only for the equator and the meridians. Most world maps don't warn that using the scale ratio to convert distances between map symbols to distances between real places almost always yields an erroneous result. Figure 2.2, for instance, would greatly inflate the distance

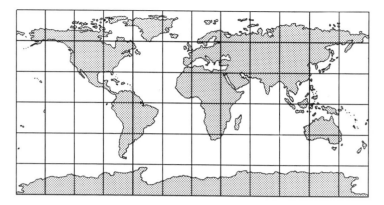

FIGURE 2.2. Equatorial cylindrical projection with true meridians.

between Chicago and Stockholm, which are far apart and both well north of the equator. Cartographers wisely avoid decorating world maps with graphic scales, which might encourage this type of abuse. In contrast, scale distortion of distance usually is negligible on large-scale maps, where the area covered is comparatively small.

Figure 2.3 helps explain the meaning and limitations of ratio scales on world maps by treating map projection as a two-stage process. Stage one shrinks the earth to a globe, for which the ratio scale is valid everywhere and in all directions. Stage two projects symbols from the globe onto a flattenable surface, such as a plane, a cone, or a cylinder, which is attached to the globe at a point or at one or two *standard lines*. On flat maps, the scale usually is constant only along these standard lines. In figure 2.2, a type of cylindrical projection called the *plane chart*, the equator is a standard line and the meridians show true scale as well.

In general, scale distortion increases with distance from the standard line. The common *developable surfaces*—plane, cone, and cylinder—allow the mapmaker to minimize distortion by centering the projection in or near the region featured on the map. World maps commonly use a cylindrical projection, centered on the equator. Figure 2.4 shows that a *secant* cylindrical projection, which cuts through the globe, yields two standard lines, whereas a *tangent* cylindrical projection, which merely touches the globe, has only one. Average distortion is less for a secant projection because the average place is closer

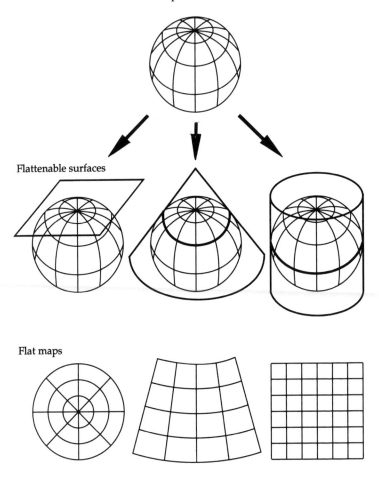

FIGURE 2.3. Developable surfaces in the second stage of map projection.

to one of the two standard lines. Conic projections are well suited to large mid-latitude areas, such as North America, Europe, and the Soviet Union, and secant conic projections offer less average distortion than tangent conic projections. *Azimuthal* projections, which use the plane as their developable surface, are used most commonly for maps of polar regions.

For each developable surface, the mapmaker can choose among a variety of projections, each with a unique pattern of distortion. Some projections, called *equivalent* or *equal-area*, allow

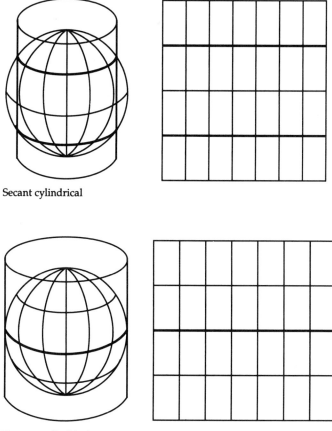

Secant cylindrical

Tangent cylindrical

FIGURE 2.4. Secant (above) and tangent (below) cylindrical projections.

the mapmaker to preserve areal relationships. Thus if South America is eight times larger than Greenland on the globe, it will also be eight times larger on an equal-area projection. Figure 2.5 shows two ways to reduce the areal distortion of the plane chart (fig. 2.2). The cylindrical equal-area projection at the top compensates for the severe poleward exaggeration by reducing the separation of the parallels as distance from the equator increases. In contrast, the sinusoidal projection below maintains true scale along the equator, all other parallels, and the central meridian and at the same time pulls the

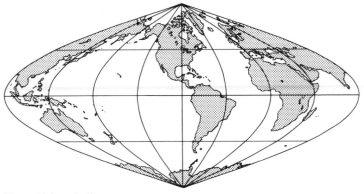

Cylindrical equal-area projection

Sinusoidal projection

FIGURE 2.5. Two varieties of equal-area cylindrical projection.

meridians inward, toward the poles, compensating for the areal exaggeration that would otherwise occur. Distortion is least pronounced in a cross-shaped zone along the equator and the central meridian and most severe between these axes toward the edge of the projection. Despite the highly distorted shapes in these "corners," the areas of continents, countries, and belts between adjoining parallels are in correct proportion.

Reduced distortion around the central meridian suggests that a sinusoidal projection "centered" on a meridian through, say, Kansas might yield a decent equal-area representation of North America, whereas a sinusoidal projection with a straight-line central meridian passing between Warsaw and Moscow would afford a suitable companion view of the Eurasian land mass. In the early 1920s, University of Chicago geography professor J. Paul Goode extended this notion of a zoned world map and devised the composite projection in figure 2.6. Goode's Interrupted Homolosine Equal-Area projection has six lobes, which join along the equator. To avoid severe pinching of the meridians toward the poles, Goode divided each lobe into two zones at about 40°—an equatorial zone based on the sinusoidal projection and a poleward zone in which the equal-area Mollweide projection portrays high-latitude areas with less east-west compression. Goode's projection mollifies the trade-off of more distorted shapes for true relative areas by giving up continuous oceans for less severely distorted land masses. If interrupted over the land to minimize distortion of the oceans, Goode's projection can be equally adept in serving studies of fisheries and other marine elements.

No flat map can match the globe in preserving areas, angles, gross shapes, distances, and directions, and any map projection is a compromise solution. Yet Goode's projection is a particularly worthy compromise when the mapmaker uses dot symbols to portray the worldwide density pattern of population, hogs, wheat, or other dryland variables. On a dot-distribution map with one dot representing 500,000 swine, for example, the spacing of these dots represents relative density. Important hog-producing regions, such as the American Midwest and northern Europe, have many closely spaced dots, whereas hog-poor regions such as India and Australia have few. But a projection that distorts area might show contrasting densities for two regions of equal size on the globe and with similar

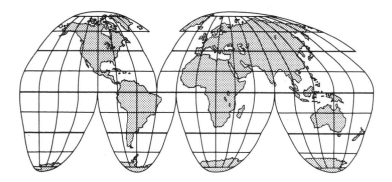

FIGURE 2.6. Goode's Homolosine Equal-Area projection.

levels of hog production; if both regions had 40 dots representing 20 million swine, the region occupying 2 cm² of the map would have a greater spacing between dots and appear less intensively involved in raising pigs than the region occupying only 1 cm². Projections that are not equal-area encourage such spurious inferences. Equivalence is also important when the map user might compare the sizes of countries or the areas covered by various map categories.

As equal-area projections preserve areas, *conformal* projections preserve local angles. That is, on a conformal projection the angle between any two intersecting lines will be the same on both globe and flat map. By compressing three-dimensional physical features onto a two-dimensional surface, a conformal projection can noticeably distort the shapes of long features, but within a small neighborhood of the point of intersection, scale will be the same in all directions and shape will be correct. Thus tiny circles on the globe remain tiny circles on a conformal map. As with all projections, though, scale still varies from place to place, and tiny circles identical in size on the globe can vary markedly in size on a conformal projection covering a large region. Although all projections distort the shapes of continents and other large territories, in general a conformal projection offers a less distorted picture of gross shape than a projection that is not conformal.

Perhaps the most striking trade-off in map projection is between conformality and equivalence. Although some projections distort both angles and areas, no projection can be both conformal and equivalent. Not only are these properties mutually exclusive, but in parts of the map well removed from the standard line(s) conformal maps severely exaggerate area and equal-area maps severely distort shape.

Two conformal projections useful in navigation illustrate how badly a map can distort area. The Mercator projection, on the left side of figure 2.7, renders Greenland as large as South America, whereas a globe would show Greenland only about one-eighth as large. North-south scale increases so sharply toward the poles that the poles themselves lie at infinity and never appear on an equatorially centered Mercator map. The right side of figure 2.7 reveals an even more severe distortion of area on the gnomonic projection, which cannot portray even half the globe.

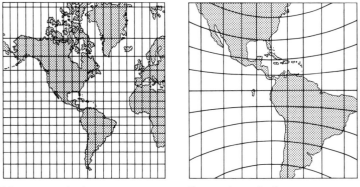

Mercator projection   Gnomonic projection

FIGURE 2.7. Straight lines on an equatorially based Mercator projection (left) are rhumb lines, which show constant geographic direction, whereas straight lines on a gnomonic projection (right) are great circles, which show the shortest route between two points.

Why, then, are these projections used at all? Although two of the worst possible perspectives for general-purpose base maps and wall maps, these maps are of enormous value to a navigator with a straightedge. On the Mercator map, for instance, a straight line is a *rhumb line* or *loxodrome*, which shows an easily followed route of constant bearing. A navigator at A can draw a straight line to B, measure with a protractor the angle between this rhumb line and the meridian, and use this bearing and a corrected compass to sail or fly from A to B. On the gnomonic map, in contrast, a straight line represents a *great circle* and shows the shortest course from A to B. An efficient navigator would identify a few intermediate points on this great-circle route, transfer these course-adjustment points from the gnomonic map to the Mercator map, mark a chain of rhumb lines between successive intermediate points, measure each rhumb line's bearing, and proceed from A to B along a compromise course of easily followed segments that collectively approximate a shortest-distance route.

Map projections distort five geographic relationships: areas, angles, gross shapes, distances, and directions. Although some projections preserve local angles but not areas, others preserve areas but not local angles. All distort large shapes noticeably (but some distort continental shapes more than others), and

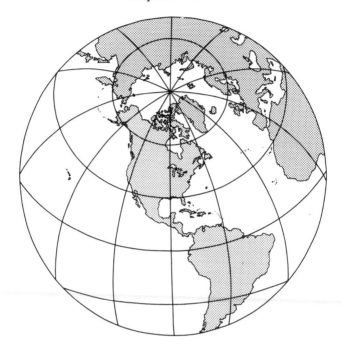

FIGURE 2.8. Oblique azimuthal equidistant projection centered on Chicago, Illinois, just east of the meridian at 90° W.

all distort at least some distances and some directions. Yet as the Mercator and gnomonic maps demonstrate, the mapmaker often can tailor the projection to serve a specific need. For instance, the oblique azimuthal *equidistant* projection in figure 2.8 shows true distance and directional relationships for shortest-distance great-circle routes converging on Chicago, Illinois. Although highly useful for someone concerned with relative proximity to Chicago, this projection is of no use for distance comparisons not involving Chicago. Moreover, its poor portrayal of the shapes and relative areas of continents, especially when extended to a full world map, limits its value as a general-purpose reference map. With an interactive computer graphics system and good mapping software, of course, map users can become their own highly versatile mapmakers and tailor projections to many unique needs.

Among the more highly tailored map projections are *cartograms*, which portray such relative measures as travel time,

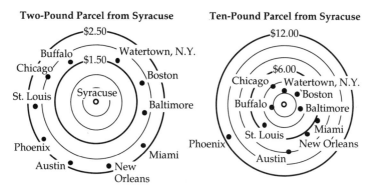

FIGURE 2.9. Distance cartograms showing relative spaces based on parcel-post rates from or to Syracuse, New York.

transport cost, and population size. Although a more conventional map might address these with tailored symbols and a standard projection, the geometry and layout of the cartogram make a strong visual statement of distance or area relationships. The distance *cartograms* in figure 2.9, for example, provide a dramatic comparison of two postal rates, which define different transport-cost spaces for their focal point, Syracuse, New York. Note that the rate for a two-pound parcel mailed to Watertown, New York, is a little more than half the rate from Syracuse to Phoenix, Arizona, whereas the corresponding rates for a ten-pound parcel more nearly reflect Watertown's relative proximity (only seventy miles north of Syracuse). These schematic maps omit boundaries and other traditional frame-of-reference features, which are less relevant here than the names of the destinations shown.

Coastlines and some national boundaries are more useful in figure 2.10, an *area cartogram*, which even includes a pseudogrid to create the visual impression of "the world on a torus." This projection is a *demographic base map,* on which the relative sizes of areal units represent population, not land area. Note that the map portrays India almost thirty times larger than Canada because the Indian population is about thirty times larger than the Canadian population, even though Canada's 3.8 million mi^2 area is much larger than India's 1.2 million mi^2. The cartogram has merged some countries with

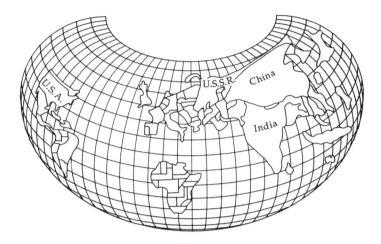

FIGURE 2.10. "World on a Torus" demographic base map is an area cartogram based on the populations of major countries.

smaller populations, demonstrating the mapmaker's political insensitivity in sacrificing nationalism for clarity. Yet traditionalist cartographers who scorn cartograms as foolish, inaccurate cartoons ignore the power of map distortions to address a wide array of communication and analytical needs.

Map Symbols

Graphic symbols complement map scale and projection by making visible the features, places, and other locational information represented on the map. By describing and differentiating features and places, map symbols serve as a graphic code for storing and retrieving data in a two-dimensional geographic framework. This code can be simple and straightforward, as on a route map drawn to show a new neighbor how to find the local elementary school; a few simple lines, labels, and Xs representing selected streets and landmarks should do. Labels such as "Elm St." and "Fire Dept." tie the map to reality and make a key or legend unnecessary.

When the purpose of the map is specific and straightforward, selection of map features also serves to suppress unimportant information. But sheet maps and atlas maps mass-produced by government mapping agencies and commercial map publishers must address a wide variety of questions, and the map's symbols must tell the user what's relevant and what isn't. Without the mapmaker present to explain unfamiliar details, these maps need a symbolic code based on an understanding of graphic logic and the limitations of visual perception. A haphazard choice of symbols, adequate for the labels and little pictures of way-finding maps and other folk cartography, can fail miserably on general-purpose maps rich in information.

Some maps, such as geologic maps and weather charts, have complex but standardized symbologies that organize an enormous amount of data meaningful only to those who understand the field and its cartographic conventions. Although as arcane to most people as a foreign language or mathematics, these maps also benefit from symbols designed according to principles of logic and communication.

Appreciating the logic of map symbols begins with understanding the three geometric categories of map symbols and the six visual variables shown in figure 2.11. Symbols on flat maps are either point symbols, line symbols, or area symbols. Road maps and most other general-purpose maps use combinations of all three: point symbols to mark the locations of landmarks and villages, line symbols to show the lengths and shapes of rivers and roads, and area symbols to depict the form and size of state parks and major cities. In contrast, *statistical maps*, which portray numerical data, commonly rely upon a single type of symbol, such as dots denoting 10,000 people or graytones representing election results by county.

Maps need contrasting symbols to portray geographic differences. As figure 2.11 illustrates, map symbols can differ in size, shape, graytone value, texture, orientation, and hue— that is, color differences as between blue, green, and red (pl. 1). Each of these six visual variables excels in portraying one kind of geographic difference. Shape, texture, and hue are effective in showing qualitative differences, as among land uses or dominant religions. For quantitative differences, size is more suited to showing variation in amount or count, such as the number of television viewers by market area, whereas graytone value is preferred for portraying differences in rate or intensity, such as the proportion of the viewing audience watching the

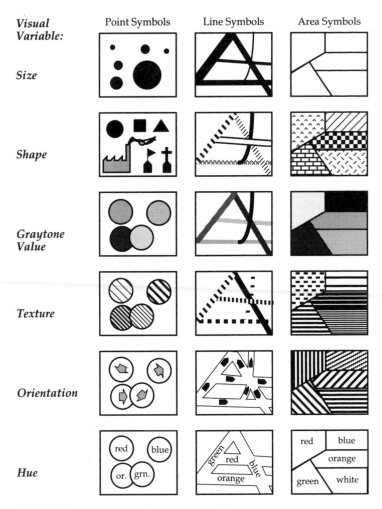

FIGURE 2.11. The six principal visual variables.

seventh game of the World Series. Symbols varying in orientation are useful mostly for representing winds, migration streams, troop movements, and other directional occurrences.

Some visual variables are unsuitable for small point symbols and thin line symbols that provide insufficient contrast with background. Hue, for instance, is more effective in showing differences in kind for area symbols than for tiny point symbols, such as the dots on a dot-distribution map. Graytone value, which usually works well in portraying percentages

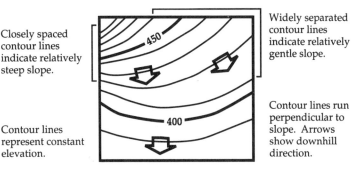

FIGURE 2.12. Elevation contours use two visual variables: spacing (texture) portrays steepness, and contour orientation is perpendicular to the direction of slope.

and rates for area symbols, is visually less effective with point and line symbols, which tend to be thinner than area symbols. Point symbols commonly rely on shape to show differences in kind and on size to show differences in amount. Line symbols usually use hue or texture to distinguish rivers from railways and town boundaries from dirt roads. Size is useful in representing magnitude for links in a network: a thick line readily suggests greater capacity or heavier traffic than a thin line implies. Area symbols usually are large enough to reveal differences in hue, graytone, and pattern, but a detail inset, with a larger scale, might be needed to show very small yet important areal units.

Some symbols combine two visual variables. For example, the elevation contours on a topographic map involve both orientation and spacing, an element of pattern. As figure 2.12 demonstrates, a contour line's direction indicates the local direction of slope because the land slopes downward perpendicular to the trend of the contour line. And the spacing of the contour lines shows the relative tilt of the land because close contours mark steep slopes and separated contours indicate gentle slopes. Similarly, the spread of dots on a dot-distribution map may show the relative sizes of hog-producing regions, whereas the spacing or clustering of these dots reveals the relative intensity and geographic concentration of production.

A poor match between the data and the visual variable can frustrate or confuse the map user. Among the worst offenders are novice mapmakers seduced by the brilliant colors of com-

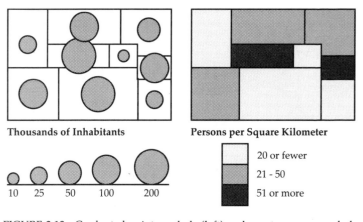

FIGURE 2.13. Graduated point symbols (left) and graytone area symbols (right) offer straightforward portrayals of population size and population density.

puter graphics systems into using reds, blues, greens, yellows, and oranges to portray quantitative differences. Contrasting hues, however visually dramatic, are not an appropriate substitute for a logical series of easily ordered graytones. Except among physicists and professional "colorists," who understand the relation between hue and wavelength of light, map users cannot easily and consistently organize colors into an ordered sequence. And those with imperfect color vision might not even distinguish reds from greens. Yet most map users can readily sort five or six graytones evenly spaced between light gray and black; decoding is simple when darker means more and lighter means less. A legend might make a bad map useful, but it can't make it efficient.

Area symbols are not the only ones useful for portraying numerical data for states, counties, and other areal units. If the map must emphasize magnitudes such as the number of inhabitants rather than intensities such as the number of persons per square mile, point symbols varying in size are more appropriate than area symbols varying in graytone. The two areal-unit maps in figure 2.13 illustrate the different graphic strategies required for portraying population size and population density. The map on the left uses *graduated point symbols* positioned near the center of each area; the size of the point symbol represents population size. At its right a *choropleth map* uses graytone symbols that fill the areal units; the relative

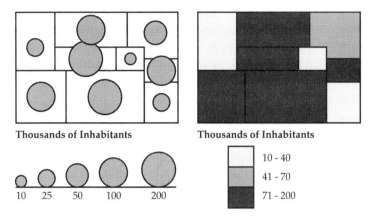

FIGURE 2.14. Map with graduated point symbols (left) using symbol size to portray magnitude demonstrates an appropriate choice of visual variable. Map with graytone area symbols (right) is ill suited to portray magnitude.

darkness of the symbol shows the concentration of population on the land.

Because the visual variables match the measures portrayed, these maps are straightforward and revealing. At the left, big point symbols represent large populations, which occur in both large and small areas, and small point symbols represent small populations. On the choropleth map to the right, a dark symbol indicates many people occupying a relatively small area, whereas a light symbol represents either relatively few people in a small area or many people spread rather thinly across a large area.

Figure 2.14 illustrates the danger of an inappropriate match between measurement and symbol. Both maps portray population size, but the choropleth map at the right is misleading because its area symbols suggest intensity, not magnitude. Note, for instance, that the dark graytone representing a large county with a large but relatively sparsely distributed population also represents a small county with an equally large but much more densely concentrated population. In contrast, the map at the left provides not only a more direct symbolic representation of population size but a clearer picture of area boundaries and area size. The map user should beware of spurious choropleth maps based on magnitude yet suggesting density or concentration.

Form and color make some map symbols easy to decode. Pictorial point symbols effectively exploit familiar forms, as when little tents represent campgrounds and tiny buildings with crosses on top indicate churches. Alphabetic symbols also use form to promote decoding, as with common abbreviations ("PO" for post office), place-names ("Baltimore"), and labels describing the type of feature ("Southern Pacific Railway"). Color conventions allow map symbols to exploit idealized associations of lakes and streams with a bright, non-murky blue and wooded areas with a wholesome, springlike green. Weather maps take advantage of perceptions of red as warm and blue as cold.

Color codes often rely more on convention than on perception, as with land-use maps, where red commonly represents retail sales and blue stands for manufacturing. Physical-political reference maps found in atlases and on schoolroom walls reinforce the convention of *hypsometric tints*, a series of color-coded elevation symbols ranging from greens to yellows to browns. Although highly useful for those who know the code, elevation tints invite misinterpretation among the unwary. The greens used to represent lowlands, for instance, might suggest lush vegetation, whereas the browns representing highlands can connote barren land—despite the many lowland deserts and highland forests throughout the world. Like map projections, map symbols can lead naive users to wrong conclusions.

Chapter 3

MAP GENERALIZATION:
LITTLE WHITE LIES AND LOTS OF THEM

A good map tells a multitude of little white lies; it suppresses truth to help the user see what needs to be seen. Reality is three-dimensional, rich in detail, and far too factual to allow a complete yet uncluttered two-dimensional graphic scale model. Indeed, a map that did not generalize would be useless. But the value of a map depends on how well its generalized geometry and generalized content reflect a chosen aspect of reality.

Geometry

Clarity demands geometric generalization because map symbols usually occupy proportionately more space on the map than the features they represent occupy on the ground. For instance, a line 1/50 inch wide representing a road on a 1:100,000-scale map is the graphic equivalent of a corridor 167 feet wide. If a road's actual right-of-way was only 40 feet wide, say, a 1/50-inch-wide line symbol would claim excess territory at scales smaller than 1:24,000. At 1:100,000, this road symbol would crowd out sidewalks, houses, lesser roads, and other features. And at still smaller scales more important features might eliminate the road itself. These more important features could include national, state, or county boundaries, which have no width whatever on the ground.

Point, line, and area symbols require different kinds of generalization. For instance, cartographers recognize the five fundamental processes of geometric line generalization described in figure 3.1. First, of course, is the *selection* of complete features for the map. Selection is a positive term that implies the suppression, or nonselection, of most features. Ideally the map author approaches selection with goals to be satisfied by

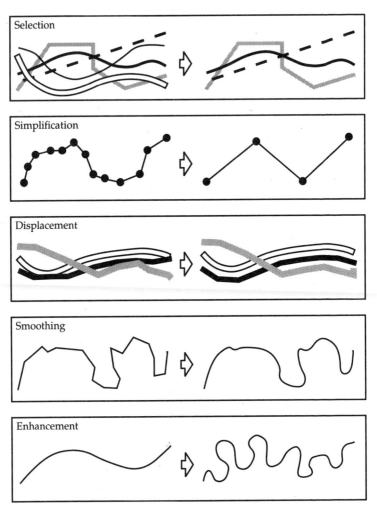

FIGURE 3.1. Elementary geometric operations in the generalization of line features.

a well-chosen subset of all possible features that might be mapped and by map symbols chosen to distinguish unlike features and provide a sense of graphic hierarchy. Features selected to support the specific theme for the map usually require more prominent symbols than background features, chosen to give a geographic frame of reference. Selecting background details that are effective in relating new information on the map to the viewer's geographic savvy and existing "mental map" often requires more insight and attention than selecting the map's main features. In the holistic process of planning a map, feature selection is the prime link between generalization and overall design.

The four remaining generalization processes in figure 3.1 alter the appearance and spatial position of linear map features represented by a series of points stored in the computer as a list of two-dimensional (X, Y) coordinates. Although the growing use of computers to generalize maps led to the isolation of these four generalization operations, traditional cartographers perform essentially the same operations by hand but with less structure, less formal awareness, and less consistency. *Simplification*, which reduces detail and angularity by eliminating points from the list, is particularly useful if excessive detail was "captured" in developing a cartographic data file, or if data developed for display at one scale are to be displayed at a smaller scale. *Displacement* avoids graphic interference by shifting apart features that otherwise would overlap or coalesce. A substantial reduction in scale, say, from 1:25,000 to 1:1,000,000, usually results in an incomprehensibly congested collection of map symbols that calls for eliminating some features and displacing others. *Smoothing*, which also diminishes detail and angularity, might displace some points and add others to the list. A prime objective of smoothing is to avoid a series of abruptly joined straight line segments. *Enhancement* adds detail to give map symbols a more realistic appearance. Lines representing streams, for instance, might be given typical meander loops, whereas shorelines might be made to look more coastlike. Enhanced map symbols are more readily interpreted as well as more aesthetic.

Point features and map labels require a somewhat different set of generalization operators. Figure 3.2 illustrates that, as with linear features, selection and displacement avoid graphic interference when too many close symbols might overlap or coalesce. When displacement moves a label ambiguously far from the feature it names, *graphic association* with a tie line or a numeric code might be needed to link the label with its symbol. *Abbreviation* is another strategy for generalizing labels on congested small-scale maps. *Aggregation* is useful where many equivalent features might overwhelm the map if ac-

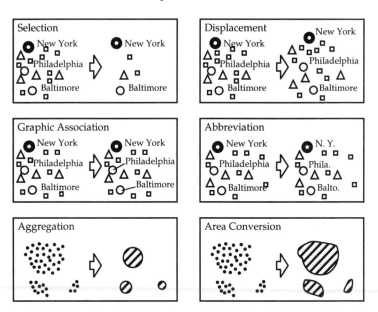

FIGURE 3.2. Elementary geometric operations in the generalization of point features and map labels.

corded separate symbols. In assigning a single symbol to several point features, as when one dot represents twenty reported tornadoes, aggregation usually requires the symbol either to portray the "center of mass" of the individual symbols it replaces or to reflect the largest of several discrete clusters.

Where scale reduction is severe, as from 1:100,000 to 1:20,000,000, *area conversion* is useful for shifting the map viewer's attention from individual occurrences of equivalent features to zones of relative concentration. For example, instead of showing individual tornadoes, the map might define a belt in which tornadoes are comparatively common. In highlighting zones of concentration or higher density, area conversion replaces all point symbols with one or more area symbols. Several density levels, perhaps labeled "severe," "moderate," and "rare," might provide a richer, less generalized geographic pattern.

Area features, as figure 3.3 demonstrates, require the largest set of generalization operators because area boundaries are

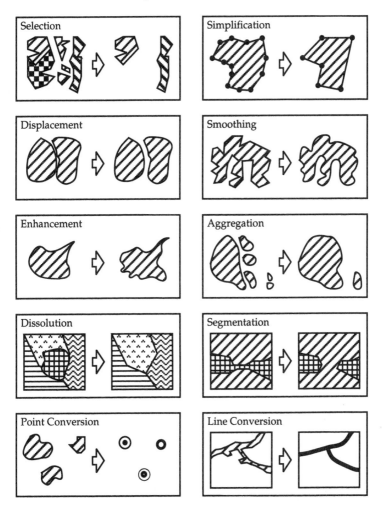

FIGURE 3.3. Elementary geometric operations in the generalization of area features.

subject to aggregation and point conversion and all five elements of line generalization as well as to several operators unique to areas. Selection is particularly important when area features must share the map with numerous linear and point features. A standardized minimum mapping size can direct the selection of area features and promote consistency among the numerous sheets of a map series. For example, 1:24,000-scale topographic maps exclude woodlands smaller than one acre unless important as landmarks or shelterbelts. Soil scientists use a less precise but equally pragmatic size threshold—the head of a pencil—to eliminate tiny, insignificant areas on soils maps.

Aggregation might override selection when a patch otherwise too small to include is either combined with one or more small, similar areas nearby or merged into a larger neighbor. On soils maps and land-use maps, which assign all land to some category, aggregation of two close but separated area features might require the *dissolution* or *segmentation* of the intervening area. A land-use map might, for example, show transportation land only for railroad yards, highway interchanges, and service areas where the right-of-way satisfies a minimum-width threshold. Simplification, displacement, smoothing, and enhancement are needed not only to refine the level of detail and to avoid graphic interference between area boundaries and other line symbols, but also to reconstruct boundaries disrupted by aggregation and segmentation.

Generalization often accommodates a substantial reduction in scale by converting area features to linear or point features. Line conversion is common on small-scale reference maps that represent all but the widest rivers with a single readily recognized line symbol of uniform width. Highway maps also help the map user by focusing not on width of right-of-way but on connectivity and orientation. In treating more compact area features as point locations, point conversion highlights large, sprawling cities such as London and Los Angeles on small-scale atlas maps and focuses the traveler's attention on highway interchanges on intermediate-scale road maps. Linear and point conversion are often necessary because an area symbol at scale would be too tiny or too thin for reliable and efficient visual identification.

Comparing two or more maps showing the same area at substantially different scales is a good way to appreciate the need for geometric generalization. Consider, for instance, the two maps in figure 3.4. The rectangles represent the same area extracted from maps published at scales of 1:24,000 and 1:250,000; enlargement of the small-scale excerpt to roughly the same size as its more detailed counterpart reveals the need for

FIGURE 3.4. Area near Northumberland, Pennsylvania, as portrayed on topographic maps at 1:24,000 (left) and 1:250,000, enlarged to roughly 1:24,000 for comparison (right).

considerable generalization at 1:250,000. The substantially fewer features shown at 1:250,000 demonstrate how feature selection helps the mapmaker avoid clutter. Note that the smaller-scale map omits most of the streets, all labels in this area, all individual buildings, and the island in the middle of the river. The railroad and highway that cross the river are smoother and farther apart, allowing space for the bridge symbols added at 1:250,000. Because the 1:24,000-scale map in a sense portrays the same area in a space over a hundred times larger, it can show many more features in much greater detail.

How precisely are symbols positioned on maps? The U.S. Office of Management and Budget addresses this concern with the National Map Accuracy Standards, honored by the U.S. Geological Survey and other federal mapping agencies. To receive the endorsement "This map complies with the National Map Accuracy Standards," a map at a scale of 1:20,000 or smaller must be checked for symbols that deviate from their correct positions by more than 1/50 inch. This tolerance reflects the limitations of surveying and mapping equipment and human hand-eye coordination. Yet only 90 percent of the points tested must meet the tolerance, and the 10 percent that don't can deviate substantially from their correct positions. Whether a failing point deviates from its true position by 2/50 inch or 20/50 inch doesn't matter—if 90 percent of the points checked meet the tolerance, the map sheet passes.

The National Map Accuracy Standards tolerate geometric generalization. Checkers test only "well-defined points" that are readily identified on the ground or on aerial photographs, easily plotted on a map, and conveniently checked for horizontal accuracy; these include survey markers, roads and railway intersections, corners of large buildings, and centers of small buildings. Guidelines encourage checkers to ignore features that might have been displaced to avoid overlap or to provide a minimum clearance between symbols exaggerated in size to ensure visibility. In areas where features are clustered, maps tend to be less accurate than in more open areas. Thus Pennsylvania villages, with comparatively narrow streets and no front yards, would yield less accurate maps than, say, Colorado villages, with wide streets, spacious front yards, and big lots. But as long as 90 percent of a sample of well-defined points not needing displacement meet the tolerance, the map sheet passes.

Maps that meet the standards show only *planimetric* distance, that is, distance measured in a plane. As figure 3.5 shows, a planimetric map compresses the three-dimensional land surface onto a two-dimensional sheet by projecting each point perpendicularly onto a horizontal plane. For two points at different elevations, the map distance between their "planimetrically accurate" positions underestimates both overland distance across the land surface and straight-line distance in three dimensions. Yet this portrayal of planimetric distance is a geometric generalization essential for large-scale flat maps.

The user should be wary, though, of the caveat "approximately positioned" or the warning "This map may not meet the National Map Accuracy Standards." In most cases such maps have been compiled from unrectified aerial photographs,

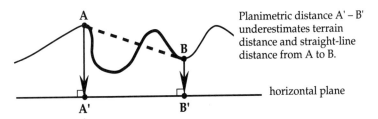

Planimetric distance A' – B' underestimates terrain distance and straight-line distance from A to B.

horizontal plane

FIGURE 3.5. Planimetric map generalizes distance by the perpendicular projection of all positions onto a horizontal plane.

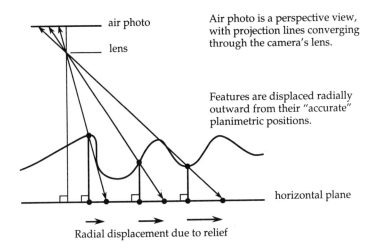

air photo

lens

Air photo is a perspective view, with projection lines converging through the camera's lens.

Features are displaced radially outward from their "accurate" planimetric positions.

horizontal plane

Radial displacement due to relief

FIGURE 3.6. A vertical aerial photograph (and any map with symbols traced directly from an air photo) is a perspective view with points displaced radially from their planimetric positions.

on which horizontal error tends to be particularly great for rugged, hilly areas. Figure 3.6 shows the difference between the air photo's perspective view of the terrain and the planimetric map's representation of distances in a horizontal plane. Because lines of sight converge through the camera's lens, the air photo displaces most points on the land surface from their planimetric positions. Note that displacement is radially outward from the center of the photo, that displacement is greater for points well above the horizontal plane than for lower points, and that displacement tends to be greater near the edges than near the center. Cartographers call this effect

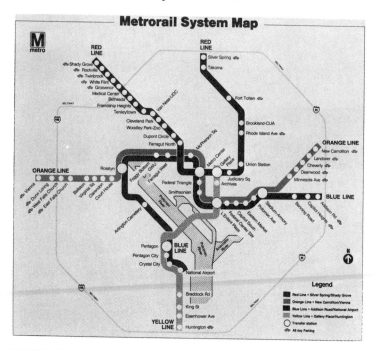

FIGURE 3.7. Linear cartogram of the Washington, D.C., Metro system.

"radial displacement due to relief," or simply *relief displacement*. An exception is the *orthophoto*, an air-photo image electronically stretched to remove relief displacement. An *orthophotomap*, produced from orthophotos, is a planimetrically accurate photo-image map.

For some maps, though, geometric accuracy is less important than linkages, adjacency, and relative position. Among the more effective highly generalized maps are the linear cartograms portraying subway and rapid transit systems. As in figure 3.7, scale is relatively large for the inner city, where the routes converge and connect; stops in the central business district might be only four or five blocks apart, and a larger scale is needed here to accommodate more route lines and station names. In contrast, toward the fringes of the city, where stations are perhaps a mile or more apart, scale can be smaller because mapped features are less dense. Contrasting colors usually differentiate the various lines; the Washington, D.C., Metro system, in fact, calls its routes the Blue Line, the

Red Line, and so forth, to enhance the effectiveness of its map. By sacrificing geometric accuracy, these schematic maps are particularly efficient in addressing the subway rider's basic questions: Where am I on the system? Where is my destination? Do I need to change trains? If so, where and to what line? In which direction do I need to go? What is the name of the station at the end of the line? How many stops do I ride before I get off? Function dictates form, and a map more "accurate" in the usual sense would not work as well.

Content

As geometric generalization seeks graphic clarity by avoiding overlapping symbols, content generalization promotes clarity of purpose or meaning by filtering out details irrelevant to the map's function or theme. Content generalization has only two essential elements, selection and classification. Selection, which serves geometric generalization by suppressing some information, promotes content generalization by choosing only relevant features. Classification, in contrast, makes the map helpfully informative as well as usable by recognizing similarities among the features chosen so that a single type of symbol can represent a group of similar features. Although all map features are in some sense unique, usually each feature cannot have a unique symbol. Even though some maps approach uniqueness by naming individual streets or numbering lots, these maps also use very few types of line symbols, to emphasize similarities among roads and property boundaries as groups. Indeed, the graphic vocabulary of most maps is limited to a small set of standardized, contrasting symbols.

Occasionally the "template effect" of standardized symbols will misinform the map user by grouping functionally different features. Standard symbols, designed for ready, unambiguous recognition and proportioned for a particular scale, are common in cartography and promote efficiency in both map production and map use. Traditional cartographers use plastic drawing templates to trace in ink the outlines of highway shields and other symbols not easily rendered freehand. Drafters can cut area and point symbols from printed sheets and stick them onto the map and can apply dashed, dotted, or parallel lines from rolls of specially printed flexible tape. Electronic publishing systems allow the mapmaker not only to choose from a menu of point, line, and area symbols provided with the software but also to design and store new forms, readily duplicated and added where needed. Consistent symbols also benefit users of the U.S. Geological Survey's series of thousands of large-scale topographic map sheets, all sharing a single graphic vocabulary. On highway maps, the key (or "legend") usually presents the complete set of symbols so that while examining the map, at least, the reader encounters no surprises. Difficulties arise, though, when a standard symbol must represent functionally dissimilar elements. Although a small typeset annotation next to the feature sometimes flags an important exception, for instance, a section of highway "under construction," mapmakers frequently omit useful warnings.

Generalized highway interchanges are a prime example of how information obscured by the template effect can mislead or inconvenience a trusting map user. The left panel of figure 3.8 is a detailed view of the interchange near Rochester, New York, between highways 104 and 590, as portrayed at 1:9,600 on a state transportation department map. Note that a motorist traveling from the east (that is, from the right) on N.Y. 104 cannot easily turn north (toward the top of the map) onto N.Y. 590. The upper right portion of the left-hand map shows that the necessary connecting lanes from N.Y. 104 were started but not completed. In contrast, the right panel shows how various commercial map publishers portray this interchange on their small-scale statewide highway maps. Two diamond-shaped interchange symbols suggest separate and equivalent connections with the eastward and westward portions of N.Y. 104. Yet the large-scale map clearly indicates that a driver expecting an easy connection from N.Y. 104 westbound onto N.Y. 590 northbound must travel to the next exit west or south and then double back. Until the road builders complete their planned connecting lanes, such discrepancies between reality and art will frustrate motorists who assume all little diamonds represent full interchanges.

Effective classification and selection often depend on a mixture of informed intuition and a good working definition. This is particularly true for geologic maps and soils maps, commonly prepared by several field scientists working in widely

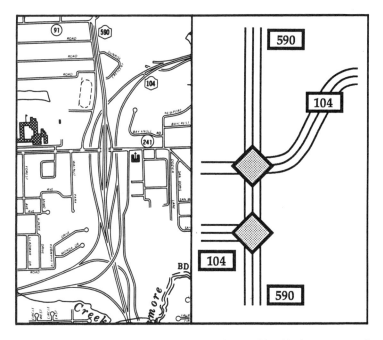

FIGURE 3.8. Highway interchange near Rochester, New York, as portrayed on a detailed transportation planning map (left) and on several commercial road maps (right).

separated places. A detailed description is necessary if two people mapping areas a hundred miles apart must identify and draw boundaries for different parts of the same feature. These descriptions should also address the mapping category's internal homogeneity and the sharpness of its "contacts" with neighboring units. In soils mapping, for instance, small patches of soil B might lie within an area labeled as soil A. This practice is accepted because these enclaves of soil B are too small to be shown separately, and because the soil scientist cannot be aware of all such enclaves. Soil mapping, after all, is slow, tedious work that requires taking samples below the surface with a drill or auger and occasionally digging a pit to examine the soil's vertical profile. Map accuracy thus depends upon the field scientist's understanding of the effects of terrain and geology (if known) on soil development as well as on expertise in selecting sample points and intuition in plotting boundaries.

That crisp, definitive lines on soils maps mark inherently fuzzy boundaries is unfortunate. More appalling, though, is the uncritical use in computerized geographic information systems of soil boundaries plotted on "unrectified" aerial photos subject to the relief-displacement error described in figure 3.6. Like quoting a public figure out of context, extracting soils data from a photomap invites misinterpretation. When placed in a database with more precise information, these data readily acquire a false aura of accuracy.

Computers generally play a positive role in map analysis and map display, the GIGO effect (garbage in, garbage out) notwithstanding. Particularly promising is the ability of computers to generalize the geometry and content of maps so that one or two geographic databases might support a broad range of display scales. Large-scale maps presenting a detailed portrayal of a small area could exploit the richness of the data, whereas computer-generalized smaller-scale displays could present a smaller selection of available features, suitably displaced to avoid graphic interference. Both the content and scale of the map can be tailored to the particular needs of individual users.

Computer-generalized maps of land use and land cover illustrate how a single database can yield radically different cartographic pictures of a landscape. The three maps in figure 3.9 show a rectangular region of approximately 700 mi^2 (1,800 km^2) that includes the city of Harrisburg, Pennsylvania, above and slightly to the right of center. A computer program generalized these maps from a large, more detailed database that represents much smaller patches of land and describes land cover with a more refined set of categories. The generalization program used different sets of weights or priorities to produce the three patterns in figure 3.9. The map at the upper left differs from the other two maps because the computer was told to emphasize urban and built-up land. This map makes some small built-up areas more visible by reducing the size of area symbols representing other land covers. In contrast, the map at the upper right reflects a high visual preference for agricultural land. A more complex set of criteria guided generalization for the display at the lower left: forest land is dominant overall, but urban land dominates agricultural land. In addition, for this lower map the computer dissolved water

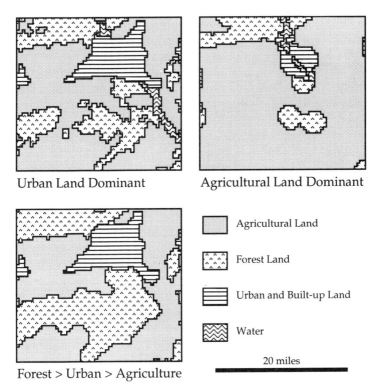

Urban Land Dominant

Agricultural Land Dominant

Forest > Urban > Agriculture

- Agricultural Land
- Forest Land
- Urban and Built-up Land
- Water

20 miles

FIGURE 3.9. Land-use and land-cover maps generalized by computer from more detailed data according to three different sets of display priorities.

areas, which were discontinuous because of variations in the width of the river. These differences in emphasis might meet the respective needs and biases of demographers, agronomists, and foresters.

Generalized maps almost always reflect judgments about the relative importance of mappable features and details. The systematic bias demonstrated by these generalized land-cover maps is not exclusive to computer-generated maps; manual cartographers have similar goals and biases, however vaguely defined and unevenly applied. Through the consistent application of explicit specifications, the computer offers the possibility of a better map. Yet whether the map's title or description reveals these biases is an important clue to the integrity of the mapmaker or publisher. Automated mapping allows ex-

perimentation with different sets of priorities. Hence computer generalization should make the cartographer more aware
of choices, values, and biases. But just because a useful and
appropriate tool is available does not mean the mapmaker
will use it. Indeed, laziness and lack of curiosity all too often
are the most important source of bias.

The choropleth map (introduced as the right-hand elements
of figs. 2.13 and 2.14) is perhaps the prime example of this bias
by default. Choropleth maps portray geographic patterns for
regions composed of areal units such as states, counties, and
voting precincts. Usually two to six graytone symbols, on a
scale from light to dark, represent two to six nonoverlapping
categories for an intensity index such as population density or
the percentage of the adult population voting in the last election. The breaks between these categories can markedly affect
the mapped pattern, and the cautious map author tests the
effects of different sets of class breaks. Mapping software can
unwittingly encourage laziness by presenting a map based
upon a "default" classification scheme that might, for instance,
divide the range of data values into five equal intervals. As a
marketing strategy, the software developer uses such default
specifications to make the product more attractive by helping
the first-time or prospective user experience success. Too
commonly, though, the naive or noncritical user accepts this
arbitrary display as the standard solution, not merely as a
starting point, and ignores the invitation of the program's
pull-down menus to explore other approaches to data classification.

Different sets of categories can lead to radically different
interpretations. The two maps in figure 3.10, for example,
offer very different impressions of the spatial pattern of homes
in the northeastern United States still lacking telephones in
1960. Both maps have three classes, portrayed with a graded
sequence of graytone area symbols that imply "low," "medium,"
and "high" rates of phonelessness. Both sets of categories use
round-number breaks, which mapmakers for some mysterious reason tend to favor. The map at the left shows a single
state, Virginia, in its high, most deficient class, and a single
state, Connecticut, in its low, most well-connected class. The
casual viewer might attribute these extremes to Virginia's higher
proportion of disadvantaged blacks and to Connecticut's af-

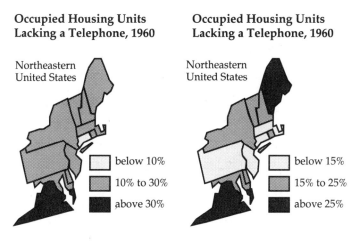

FIGURE 3.10. Different sets of class breaks applied to the same data yield different-looking choropleth maps.

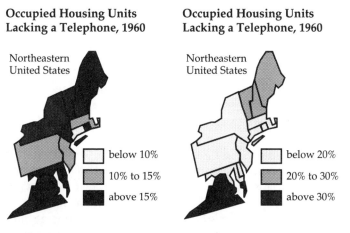

FIGURE 3.11. Class breaks can be manipulated to yield choropleth maps supporting politically divergent interpretations.

fluent suburbs and regard the remaining states as homogeneously "average." In contrast, the map at the right portrays a more balanced distribution of states among the three groups and suggests a different interpretation. Both states in the high category have substantial dispersed rural populations, and all four in the low category are highly urban and industrialized.

Moreover, a smaller middle group suggests less overall homogeneity.

Machiavellian bias can easily manipulate the message of a choropleth map. Figure 3.11, for example, presents two cartographic treatments with substantially different political interpretations. The map on the left uses rounded breaks at 10 percent and 15 percent, forcing most states into its high, poorly connected category and suggesting a Northeast with generally poor communications. Perhaps the government is ineffective in regulating a gouging telecommunications industry or in eradicating poverty. Its counterpart on the right uses rounded breaks at 20 percent and 30 percent to paint a rosier picture, with only one state in the high group and eight in the low, well-served category. Perhaps government regulation is effective, industry benign, and poverty rare.

The four maps in figures 3.10 and 3.11 hold two lessons for the skeptical map reader. First, a single choropleth map presents only one of many possible views of a geographic variable. And second, the white lies of map generalization might also mask the real lies of the political propagandist.

Intuition and Ethics in Map Generalization

Small-scale generalized maps often are authored views of a landscape or a set of spatial data. Like the author of any scholarly work or artistic creation based on reality, the conscientious map author not only examines a variety of sources but relies on extensive experience with the information or region portrayed. Intuition and induction guide the choice of features, graphic hierarchy, and abstraction of detail. The map is as it is because the map author "knows" how it should look. This knowledge, of course, might be faulty, or the resulting graphic interpretation might differ significantly from that of another competent observer. As is often the case, two views might both be valid.

Chapter 4

BLUNDERS THAT MISLEAD

Some maps fail because of the mapmaker's ignorance or oversight. The range of blunders affecting maps includes graphic scales that invite users to estimate distances from world maps, maps based on incompatible sources, misspelled place-names, and graytone symbols changed by poor printing or poor planning. By definition a blunder is not a lie, but the informed map user must be aware of cartographic fallibility, and even of a bit of mischief.

Cartographic Carelessness

Mapmakers are human, and they make mistakes. Although poor training and sloppy design account for some errors, most cartographic blunders reflect a combination of inattention and inadequate editing. If the mapmaker is rushed, if the employer views willingness to work for minimal wages as more important than skill in doing the job, or if no one checks and rechecks the work, missing or misplaced features and misspelled labels are inevitable.

Large-scale base maps have surprisingly few errors. A costly but efficient bureaucratic structure at government mapping agencies usually guarantees a highly accurate product. Several layers of fact checking and editing support technicians or contractors selected for skill and concern with quality. Making topographic maps is a somewhat tedious, multistep manufacturing process, and using outside contractors for compilation or drafting requires a strong commitment to quality control buttressed by the bureaucrat's inherent fear of embarrassment. Blunders occasionally slip through, but these are rare.

Errors are more common on derivative maps—that is, maps compiled from other maps—than on basic maps compiled from air photos and other primary data. Artists lacking carto-graphic training and an appreciation of geographic details draw most tourist maps and news maps, and poorly paid drafting technicians produce most American street maps. Omission and garbling are particularly likely when information is transferred manually from one map to another. Getting all the appropriate information from the large-scale base map onto a small-scale derivative map is not an easy chore. Sever-al base maps might be needed, the compiler might not have a clear idea of what is necessary, or several compilers might work on the same map. Using another derivative map as a source can save time, but only at the risk of incorporating someone else's errors.

Map blunders make amusing anecdotes, and the press helps keep cartographers conscientious by reporting the more out-rageous ones. In the early 1960s, for instance, the American Automobile Association "lost Seattle," as the Associated Press reported the accidental omission of the country's twenty-third largest city from the AAA's United States road map. Embar-rassed, the AAA confessed that "it just fell through the editing crack" and ordered an expensive recall and reprinting.

Equally mortified was the Canadian government tourist office that omitted Ottawa from an airline map in a brochure prepared to attract British tourists. The official explanation that Ottawa had not been a major international point of entry and that the map was compiled before initiation of direct New York–Ottawa air service didn't diminish the irritation of Ottawa residents. The map included Calgary, Regina, and Winnipeg, and as an executive of the city's Capital Visitors and Conven-tion Bureau noted, "Ottawa should be shown in any case, even if the only point of entry was by two-man kayak."

Faulty map reading almost led to an international incident in 1988, when the Manila press reported the Malaysian annex-ation of the Turtle Islands. News maps showing the Malay-sian encroachment supported three days of media hysteria and saber rattling. These maps were later traced to the erroneous reading of an American navigation chart by a Philippine naval officer who mistook a line representing the recommended deepwater route for ships passing the Turtle Islands for the boundary of Malaysia's newly declared exclusive economic zone.

Although map blunders provoking outrage between minor powers make amusing anecdotes, inaccurate maps in a war zone can be deadly. The American Civil War illustrates the effect on both sides of wildly conflicting topographic maps and inadequate numbers of trained topographic engineers and geographers. In 1862, for instance, the Union army planned a swift defeat of the Confederates by capturing their capital, Richmond. But unexpected obstacles slowed the northern army's advance after General McClellan's staff based battle plans on inaccurate maps. A lack of good maps also plagued Confederate forces, who were unaware of strategic advantages that would have allowed them to overwhelm McClellan's retreating army.

Modern warfare is particularly vulnerable to bad maps, as the 1983 invasion of Grenada by United States troops and their Caribbean allies demonstrates. The only cartographic intelligence distributed to troops carrying out this politically convenient rescue of American medical students consisted of hastily printed copies of a few obsolete British maps and a tourist map with a military grid added. An air attack destroyed a mental hospital not marked on the maps. Another air strike, ordered by a field commander using one set of grid coordinates but carried out by planes using a map with another grid, wounded eighteen soldiers, one fatally.

Journalists' and social scientists' accounts of the invasion added further cartographic insult. In addition to misplaced symbols and misspelled place-names, one group of coauthors (or their free-lance illustrator) distorted the size and relative position of the Grenadian state's two smaller members, the islands of Carriacou and Petite Martinique. As the lower right part of figure 4.1 illustrates, the book's regional locator map made these islands much smaller than their true relative sizes and moved them much closer to the main island. These errors probably originated in careless compilation from an official map sheet on which an inset showed the smaller islands at a smaller scale than the main island. Since an earlier derived map had the same errors, the authors apparently based their map on a faulty source, which lax editorial checking obviously failed to detect.

The news media have their own cartographic glitches. Errors on news maps reflect both the minimal cartographic

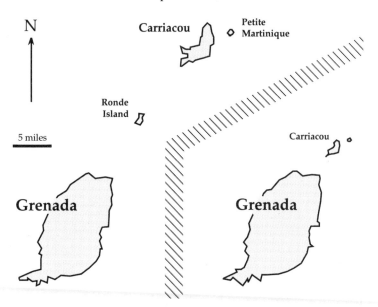

FIGURE 4.1. Reconstruction of a journalistic misrepresentation (right) of the size and location of Grenada's smaller islands, as presented in a book published shortly after the 1983 invasion. The left and upper portion of the figure portrays Carriacou and Petite Martinique correctly.

knowledge of most newspaper artists and the high-pressure, deadline-driven environment in which news is gathered, processed, and published. The use of graphics in newspapers began to grow in the late nineteenth century, when photoengraving enabled news publishers to use more line drawings and photographs and feature syndicates arose to serve smaller papers unable to hire their own artists. Unfortunately, technological advances that made news maps less expensive and more common also allowed poorly skilled, geographically illiterate artists to make publishable maps rapidly. Many news maps distributed about 1900 had the crude boundaries and careless spelling exemplified by figure 4.2, a map accompanying a syndicated story about a coming eclipse; note the extra *n* in the label for Illinois. Hastily drawn news maps have also annexed Michigan's northern peninsula to Wisconsin, Virginia's eastern counties on the Delmarva Peninsula into Maryland, and both North Korea and South Korea (and part of China) to the Soviet Union.

PATH OF THE TOTAL ECLIPSE.

FIGURE 4.2. On 28 May 1900 the *Cortland* (N.Y.) *Evening Standard* printed this hand-drawn map, received from a feature syndicate with the name Illinois misspelled.

Computer graphics, a more recent impetus for journalistic cartography, makes it easier for reporters and editors to alter decent-looking base maps and inadvertently eliminate features and misplace symbols and labels. Typical examples include adding Finnish territory to the Soviet Union on the map decorating a *New York Times* article on Canadian–Soviet relations and switching the labels identifying New Hampshire and Vermont on a Knight-Ridder Graphics Network map of areas in the United States affected by drought.

Perhaps more annoying are blunders on road maps and street maps, particularly when the place you are trying to find is missing, misplaced, misindexed, mislabeled, or badly mis-shaped. That these errors are not more common is surprising, though. Publishers of street and highway maps must manage a complex, constantly changing database and produce a low-cost, enormously detailed product for largely unappreciative consumers in a highly competitive market. Perhaps because oil companies distributed free road maps for several decades until the early 1970s, and because state and local tourist councils perpetuate the free travel map, the American map buyer has little appreciation of the well-designed, highly accurate maps that the European map user is conditioned to respect, demand, and pay for.

Blunders on street maps reflect how the maps are made. Basic data for established parts of an area can be found on large-scale topographic maps published by the Geological Survey. These maps are in the public domain and can be copied freely, but their publication scale of 1:24,000 allows room for few street names, and many map sheets are ten years or more out of date. Mapmakers thus turn to maps maintained by city and county engineering and highway departments for street names as well as for new streets and other changes. Personnel responsible for copying street alignments, typeset-ting, and type placement are sometimes inexperienced or inat-tentive, and editing is not always thorough. A file of custom-ers' complaints can be a useful though not fully reliable source of corrections for the next edition. Some publishers send drafts of their maps to planning and engineering departments for com-ment and unrealistically depend on overworked civil servants for additional editing.

Unfortunately, many official city maps show the rights-of-way of approved streets that were never cleared, graded, or paved. Often a planned but otherwise fictitious street persists on the city engineer's map until it is officially deleted, as when a developer petitions the planning board to build across the right-of-way or a homeowner attempts to buy the adjoin-ing strip of land. Street-map publishers who compile only from municipal maps are likely to pick up a number of "paper streets" like Garden Street and Pinnacle Street in figure 4.3, found on an official map for Syracuse, New York. Geological Survey and New York State quadrangle maps of the area don't show these streets, but the mapmaker might conveniently

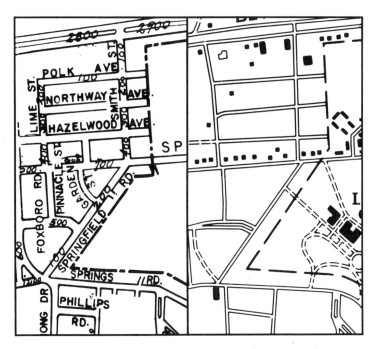

FIGURE 4.3. Two paper streets in Syracuse, New York, as shown on a municipal street map (left panel; see Garden Street and Pinnacle Street just to the left of the center of the map) and the corresponding area as represented on a New York State highway planning map. (In the left panel a tiny arrow points to block-long Pinnacle Street, which is parallel to Smith Street.)

assume that the "official" map provided by local government is more likely to be accurate than a map from Washington or the state capital. A slim profit margin usually precludes field checking, and up-to-date large-scale air photos are expensive to purchase and seldom convenient to consult. Because a mapmaker in a distant city cannot readily determine whether the feature is an obsolete paper street or a recently opened thoroughfare, commercial street maps occasionally pick up phantom streets.

Deliberate Blunders

Although none dare talk about it, publishers of street maps also turn to each other for street names and changes. The

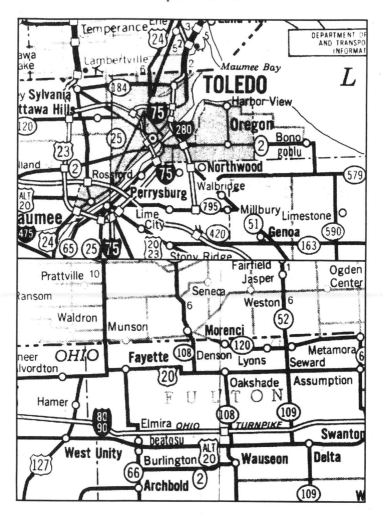

FIGURE 4.4. Fictitious towns "goblu" (above, to the right, below Bono) and "beatosu" (toward the bottom, above Burlington) on the 1979 Michigan highway map reflect an unknown mapmaker's support for the University of Michigan football team (the Blue) over its traditional rival, Ohio State University (OSU).

euphemism for this type of compilation is "editing the competition," but the legal term is copyright infringement—if you crib from a single source and get caught. To be able to demonstrate copyright infringement in court, and possibly to enjoy a cash settlement by catching a careless competitor in the act, map publishers have been known to deliberately falsify their maps by adding "trap streets." As deterrents to the theft of copyright-protected information, trap streets are usually placed subtly, in out-of-way locations unlikely to confuse or antagonize map users. Map publishers are understandably reluctant to talk about this dubious practice of deliberate falsification.

Map drafters having fun are another source of cartographic fiction. Michigan's state highway map for 1979, for instance, included two fictitious towns reflecting the traditional football rivalry between the University of Michigan and Ohio State University. As figure 4.4 shows, the cartographic culprit clearly was a not only a Michigan fan but a loyal citizen of Michigan, perpetrating place-name pollution only in neighboring portions of Ohio—and perhaps thinking the editor would be less careful in checking out-of-state features. Toledo's new eastern suburb "goblu" is a slightly compacted version of the familiar cheer of fans rooting for the Michigan Blue (a nickname based on the primary school color), and the new town "beatosu," north of Burlington and south of the Ohio Turnpike, reflects the Blue's principal annual gridiron goal, defeating OSU.

A more personal example of creative cartography is Mount Richard, which in the early 1970s suddenly appeared on the continental divide on a county map prepared in Boulder, Colorado. Believed to be the work of Richard Ciacci, a draftsman in the public works department, Mount Richard was not discovered for two years. Such pranks raise questions about the extent of yet-undetected mischief by mapmakers reaching for geographic immortality.

Distorted Graytones:
Not Getting What the Map Author Saw

Printing can radically alter the appearance of a map, and failure to plan for the distorting effects of reproduction can yield thousands of printed maps that look quite different from the original artwork. Graytones can be particularly fragile during two steps in map reproduction: the photographic transfer of the map image from a drawing or computer plot onto a photographic negative (or in some cases directly onto a press plate) and the transfer of the inked image from the printing plate onto the paper. Overexposure or underdevelopment of the photograph or an underinking printing press yields a faint image, from which fine type or fine area patterns might have disappeared. Underexposure or overdevelopment of the photograph or an overinking press might fill in the corners and tightly closed loops of small type and can noticeably darken some graytone area symbols. Usually the culprit is an overinking printing press, which fattens image elements through a malfunction called *ink spread*. Photographic exposure and development can be controlled, and if necessary the artwork can be reshot. But at least some overinking is common for a significant part of most printing runs, and in many cases significant overinking occurs throughout the run.

Fine dot screens used as area symbols are vulnerable to ink spread, and the map designer who underestimates the effect of ink spread on screens with dots spaced 120 or more to the inch (47 or more per centimeter) risks medium grays that turn out black and choropleth maps on which low becomes high and high becomes low. Figure 4.5 uses greatly enlarged views of two hypothetical graytone area symbols to illustrate this effect on a relatively fine 150-line screen (59 lines per centimeter), with dots spaced 0.007 inch (0.169 mm) apart. In this simulation a small amount of ink spread that increases the radius of each dot by 0.001 inch (0.025 mm) raises the black area of a 20 percent screen to 49 percent and that of an 80 percent screen to 96 percent. Screens for original graytones of less than 50 percent black grow darker because ink added around the edges of tiny black dots makes the dots larger, in some cases causing them to coalesce. Screens for original graytones of more than 50 percent black grow darker because ink added on the inside edges of tiny clear dots on a black background makes the clear dots smaller, in some cases completely filling them in. The cautious mapmaker adjusts screen texture to printing quality. The same amount of ink spread applied to moderately coarse, 65-line screens (26 lines per centimeter) would increase the 20 percent screen to only 32 percent black and the 80 percent screen to only 89 percent black.

The map user should be particularly wary of choropleth maps with both fine and coarse dot screens. Before examining the mapped patterns for clusters of dark, high-value areas and light, low-value areas, the user should inspect the key for possible inversion of part of the graytone sequence. Ink spread might, for example, have reproduced a 40 percent fine graytone as darker than a comparatively coarse 60 percent graytone. Overinking might also have aggregated the two or three highest categories into a single group represented by solid black area symbols. Ink spread can produce particularly troublesome distortions of visual contrast and graphic logic on maps printed in color.

Laser printers and other remote electronic displays are another source of graphic noise. Pattern description codes that produce one symbol on the map designer's computer might yield a very different area symbol when the map is printed or displayed on the other side of the room or, if transmitted in compact "object code" over a telecommunications network, on the other side of the continent or world. A graphics

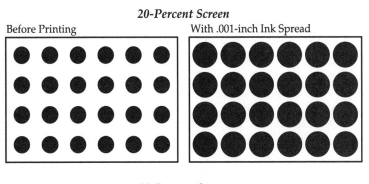

20-Percent Screen

Before Printing   With .001-inch Ink Spread

80-Percent Screen

Before Printing   With .001-inch Ink Spread

FIGURE 4.5. Enlarged diagram showing the effects of 0.001 inch of ink spread on 20 percent (above) and 80 percent (below) 150-line graytone dot screens.

workstation with a moderate-resolution monitor might display a graytone area symbol at 36 dots per inch (14 dots per centimeter), for instance, whereas a distant laser printer programmed to take full advantage of its higher resolution might render the same area symbol with a much finer, 150-line dot screen. A map author unable to review a laser-printer image before it is printed could be surprised and embarrassed by the effect of ink spread on a graphically unstable fine-dot screen.

Temporal Inconsistency:
What a Difference a Day (or Year or Decade) Makes

Maps are like milk: their information is perishable, and it is wise to check the date. But even when the map author provides one, the date might reflect the time of publication, not the time for which the information was gathered. And when the map was compiled from more than one source or through a long, tedious field survey, the information itself might be so temporally variable as to require not a single date but a range of dates. Particularly troublesome is the carefully dated or current-situation map that shows obsolete features or omits more recent ones. These errors might be few and not readily apparent; a map that is 99.9 percent accurate easily deceives most users.

Inaccurately dated or temporally inconsistent maps can be a particular hazard when the information portrayed is volatile. A map of past geological conditions might inaccurately estimate its date by thousands or even millions of years and still be useful, whereas the temperature and pressure observations used to prepare weather maps must be synchronized to within an hour or less. Moreover, maps forecasting weather patterns must state accurately the date and time for the forecast.

Historians in particular should be skeptical of dates on maps. Medieval maps, for instance, can cover a much broader range of time than a single year-date suggests. As historian of cartography David Woodward has noted, medieval *mappaemundi* (world maps), instead of providing an accurate or perceived image of the earth at an instant of time, often "consist of historical aggregations or cumulative inventories of events that occur in space." For instance, the famous Hereford map, named for the British cathedral that owns it, was compiled about 1290 from a variety of sources. Its place-names present an asynchronous geography ranging from the fourth-century Roman Empire to contemporary thirteenth-century England.

Conscientious users of modern maps read whatever fine print the mapmaker provides. A large-scale topographic map released in year N with a publication date of year N – 2, for instance, might be based on air photos taken in year N – 3 or N – 4, and might have been field checked in year N – 2 or N – 3. But field checking might not detect all significant changes, and maps of areas undergoing rapid urban development often are appallingly obsolete. Moreover, derived maps without fine print can be the cartographic equivalent of snake oil. Because of publication delays and slow revision, "new" de-rived maps may well be ten years or more out of date. Or they may be only four years out of date in some areas, and ten years or more in others.

A temporally accurate map need not focus on a single date or include only features that currently exist or once existed. Planners, for instance, need maps to record their previous decisions about future projects so that new decisions do not conflict with old ones. Fairfax County, Virginia, learned that inaccurate planning maps can have costly consequences. Be-cause of inadequate mapping, a developer was authorized to start building a subdivision in the path of a planned limited-access highway. Buying the seventeen affected lots cost coun-ty taxpayers $1.5 million. The same highway also required the county to buy and raze five new homes in another subdivision. After these two embarrassing incidents, county officials set up a special mapping office.

Temporal consistency is also troublesome for users of sta-tistical maps produced from census and survey data. Choro-pleth maps and other data maps sometimes portray a ratio or other index that compares data collected for different instants or periods of time. Usually the temporal incompatibility is minor, as when the Bureau of the Census computes per capita income by dividing an area's total income for the calendar year 1989 by its population as enumerated on 1 April 1990—getting an accurate count of the population and honest and reliable estimates of personal income is far more problematic. One must beware, though, of indexes that relate mid-decade survey data and beginning-of-decade census tabulations, for

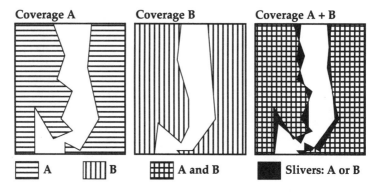

FIGURE 4.6. Spurious sliver polygons result from the overlay of two inaccurately digitized coverages A and B.

example, by dividing 1995 income totals by 1990 populations. Migration can significantly distort ratios based on asynchronous data, particularly for small areal units such as suburban towns, which can grow by 500 percent or more in ten years.

International data based on inconsistent definitions as well as asynchronous censuses or surveys can yield highly questionable maps. Worldwide maps of poverty, occupational categories, and the proportion of the population living in urban areas are inherently imprecise because of significant international differences in the relevant definitions. World maps based on statistical data are particularly suspect—whatever validity they have arises not only from knowing adjustment by scholars or United Nations officials but from broad, very general categories that tend to mask spurious differences within the groups of more and less developed nations.

Maps based on electronic data files can be highly erroneous, especially when several sources contributed the data and the user or compiler lacked the time or interest to verify their accuracy. Obviously inaccurate data from careless, profit-driven firms have unpleasantly surprised purchasers of street network information, and even data from reliable vendors occasionally have infuriating errors. Without careful editing, streets are more easily omitted or misplaced in a computer database than on a paper street map. Transfer from paper map to electronic format invites many inconsistencies, especially when many sets of features, or "coverages," can be related in a geographic information system. Roads and boundaries entered from adjoining map sheets often fail to align at the common edge, and features extracted from different maps of the same area might be misaligned, perhaps because their source maps have different projections. Overlays of spatially or temporally incompatible data—for example, the overlay of two closely related coverages collected separately— can yield slivers and other spurious polygons, as the example in figure 4.6 demonstrates. Moreover, the software used to process the information might also be flawed. Although software errors can be blatant, the possibility of subtle programming errors that could lead to disastrous decisions should encourage the user to carefully explore unfamiliar data and software. "Garbage in, garbage out" is a useful warning, but sometimes you can't tell the data are garbage until they have been used for a while.

Chapter 5

MAPS THAT ADVERTISE

What do advertising and cartography have in common? Without doubt the best answer is their shared need to communicate a limited version of the truth. An advertisement must create an image that's appealing and a map must present an image that's clear, but neither can meet its goal by telling or showing everything. In promoting a favorable comparison with similar products, differentiating a product from its competitors, or flattering a corporate image, an ad must suppress or play down the presence of salt and saturated fat, a poor frequency-of-repair record, or convictions for violating antitrust, fair-employment, and environmental regulations. Likewise, the map must omit details that would confuse or distract.

When the product or service involves location or place, the ad often includes a map, sometimes prominently. Two mildly Machiavellian motives account for the use of maps as central elements in some advertising campaigns and as important props in others. First, as art directors and marketing specialists have discovered, the map's need to avoid graphic interference can serve the advertiser's need to suppress and exaggerate. Indeed, maps that advertise tend to be more generalized than graphic clarity demands. Second, ads must attract attention, and maps are proven attention getters. In advertising, maps that decorate seem at least as common as maps that inform.

This chapter examines cartographic distortion in commercial advertising. It looks first at transportation ads, which frequently employ maps to exaggerate the quality of service. Occasionally the distortion is deliberately overdone to make the map a graphic pun that uses artistic flair to make the advertiser's point about convenience or improved service. The second section shows how a map touting one or a few locations can convey an image of convenience or exclusiveness. The final section treats maps that promote chain stores and franchise businesses by equating numerousness with success and quality. Simple scenarios illustrate how advertisers and ad agencies exploit maps as marketing tools. The contrived levity of the examples discussed reflects real advertisers' reluctance to submit to public criticism as well as the lighthearted, half-serious attitude of many ad maps.

Transport Ads:
Gentle Lines and Well-Connected Cities

It is 1875. As president of the soon-to-be-completed Helter, Skelter, and Northern Railway, you need to advertise your route both in a timetable and in the shippers' bible, the *Official Guide of the Railways*. Your engineering department's small-scale map (fig. 5.1) seems not quite suitable. No point in drawing attention to your principal competitor, the Helter, Skelter and Yon, whose more direct route actually reaches

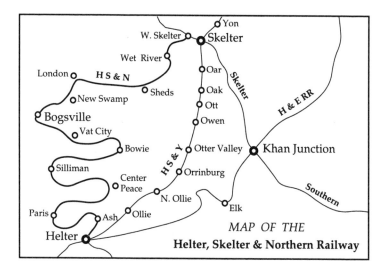

FIGURE 5.1. Engineering department's map of the Helter, Skelter and Northern Railway.

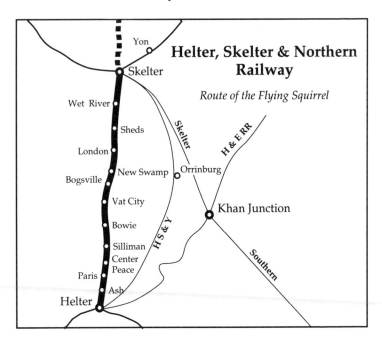

FIGURE 5.2. Advertising and timetable map of the Helter, Skelter and Northern Railway.

downtown Skelter rather than terminating in West Skelter, three miles from the Skelter business district. Nor would you want to publicize the HS&N's tortuous trek up the floodplain of Catfish Creek to the wishful city of Bogsville, which willingly bought construction bonds to attract a railroad. And since you seek further capital, the map might also add credence to the still dubious promise "and Northern" in the company name. Further, the overall shape and geometry of the engineer's map seems inappropriate: you want the HS&N clearly in the center, and you need room to show the names of as many cities, towns, and hamlets as you dare claim for your sparsely populated right-of-way.

You explain your needs to a free-lance cartographer with experience in railway advertising, who three days later proudly delivers the map in figure 5.2. Amazed, pleased, and grateful, you first stare at the map for a minute and wonder whether it shows the correct railroad. Its dominant visual feature is a nearly straight line connecting Helter with Skelter. Having captured all towns and hamlets close enough to warrant a crossroads sign or unoccupied shelter, your cartographic fiction shrewdly suppresses the label "West Skelter" and suggests a direct link with other rail lines in Skelter. (If the naive viewer wants to assume a common connection downtown, well, that's his inference.) And a nearly-as-prominent dashed extension running off the map to the north suggests imminent new construction and vast possibilities for investment and profit. In contrast, a thin, graphically weak line portrays your principal competitor, the Helter, Skelter and Yon Railroad, as not only offering a more devious route between Helter and Skelter but also nearly bypassing Yon. Thanks to cartographic license, the Helter, Skelter and Northern Railway has become an attractive option for distant shippers and distant investors.

Time flies, and it is now the turn of the twenty-first century. In a new reincarnation, you are president of the recently formed Upward Airlines. And once again you need a map for schedules and media advertising. Yet the cartographic challenge has changed: instead of deemphasizing a sinuous route and ignoring a single parallel competitor, you now need to draw attention to the number of cities served and the overall integration of the Upward Airlines system. Like other national airlines, Upward operates a hub-and-spoke system, with planes converging several times a day on each of its two hubs, where passengers get off, hike to another gate, and board a second aircraft.

Planes fly fairly direct routes, as everyone knows, so your modern-day advertising agency freely sacrifices straight lines and their impression of geographic directness for the dramatic, rather busy-looking map in figure 5.3. Cartographic license lies largely in the map's suggestion that all flights on each spoke are nonstop. In reality, though, all but one of the flights from Burlington, Vermont, land at Buffalo, New York, on their way to the Saint Louis hub, and similar intermediate stops occur on other spokes in the system. Moreover, service is not equally frequent along all the spokes. For instance, Bismarck, North Dakota, has only one plane a day in and out, and then only on weekdays. And if fewer than 20 percent of its seats are sold, the midday service from Denver to Shreveport, Louisiana—but rarely the inward flight toward the hub—is canceled because of "mechanical problems."

Upward Airlines—Reach for the Clouds

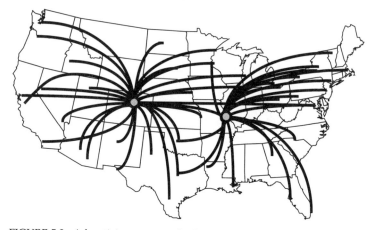

FIGURE 5.3. Advertising map emphasizing Upward Airlines' service area and connections.

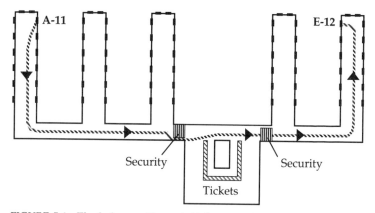

FIGURE 5.4. The hub map Upward Airlines won't publicize: the fifteen-minute luggage-laden trek from gate A-11 to gate E-12, with a stop at the security checkpoint.

Another expediently misleading generalization is the map's suggestion that connections at Upward's hubs are convenient. In addition to frequent, frustrating delays on taxiways and in crowded boarding lounges and to less frequent but more frustrating overnight stays because of missed late-evening con-

Go island-hopping with Transcendental Airlines

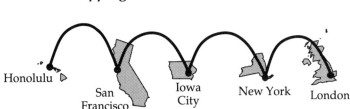

Honolulu

San Francisco

Iowa City

New York

London

FIGURE 5.5. Catchy map useful for airline ads in Iowa City newspapers.

nections, the hub symbol tells the prospective passenger nothing about the likelihood of a quarter-mile walk with infant, garment bag, or other "carry-on items" from gate A-11 to gate E-12. Formed by the merger of Westward Airlines and Northeastward Airlines, Upward Airlines must make do with its predecessors' gates at opposite ends of the terminal. Thus figure 5.4 is the very last map likely to appear in one of Upward's ads. Of course, a small, nonhub competitor offering direct service between two cities linked through Upward's hub might exploit such a map in its own advertising campaign!

Advertising agencies serving airlines can have great fun with maps if the client has a sense of humor. In addition to decorating maps with pictograms of impressive skyscrapers, museums, golfers, girls in bathing suits, and other symbols of culture or leisure, graphic designers can create a variety of cartographic puns by manipulating maplike images. As figure 5.5 demonstrates, ad maps can be effective eye-catchers as well as pandering to local pride. After all, people in Iowa City, Iowa, like to see themselves considered on a level with London, New York, and San Francisco.

Enticement and Accessibility:
Ad Maps with a Single-Place Focus

In an ad promoting a store, resort, or other business, a map might not only offer travel directions but also stimulate demand. For many goods and services, the trip itself is an important part of the purchase. If bad roads, walking through or parking in an unsafe neighborhood, or heavy traffic would make the trip an ordeal, the buyer might reasonably consider travel an added cost and look for a less expensive alternative. Thus a map promoting a retail outlet needs not only to suggest straightforward routes for getting there but also to present an image of convenient accessibility. And if an attractive image requires distorted distances and directions, the advertising map will indeed distort. Accuracy and precision, after all, are seldom prime goals in advertising.

Accessibility is particularly important for products needed in a hurry. And few customers feel as much duress as the do-it-yourself plumber confronting a sudden shower in the basement or a running toilet that tinkering has only made worse. Thus when Rudy Swenson stops by your newly opened advertising firm to discuss a campaign for his plumbing-supplies house, you immediately impress him by suggesting a map as the ideal display ad and the classified telephone book as its ideal vehicle. Like other shoppers, after all, do-it-yourself plumbers are conditioned to respond to emergencies by looking in the yellow pages. (Indeed, that the classified phone directory lists licensed plumbers offers an added sense of security.)

Rudy's business is ripe for a tailor-made ad map. None of his few competitors in town are on the same street, so your map need not give circuitous directions to avoid a business rival. Although his store is in the low-rent section of the city, the two streets intersecting at the corner extend far enough out of town to avoid the immediate area's reputation for vice and mayhem. (A trip to this area might even appeal to the do-it-yourself plumber's penchant for living on the edge.) Although you briefly consider sprinkling some imaginative travel-time estimates across the map, you decide instead to invoke the advertising cartographer's ever-reliable ploy—place-name dropping. Your finished design (fig. 5.6) shamelessly mentions a host of neighboring towns, some of which have their own plumbing-parts dealers. Yet the thought of customers traveling thirty miles over State Route 10 from Canton and a good forty-five miles over tortuous State Route 19 from Cynwyd pleases Rudy as much as it impresses local homeowners. And there near the center of the map is a drawing of Rudy's building, its perspective seeming to justify your wild and willful distortion of the region's geography. (What newcomer would guess that East Hills is actually closer to Rudy's than Westvale? But

Rudy's Plumbing Supplies: as close as you are

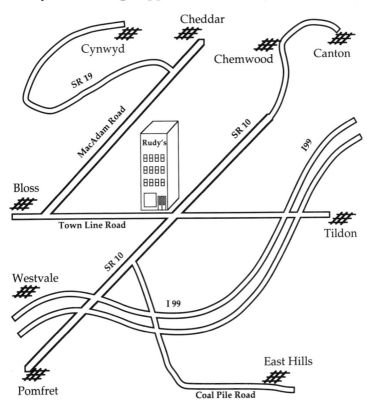

FIGURE 5.6. Map used in classified telephone directory display ad for Rudy's Plumbing Supplies does more than give directions.

rather than take offense, residents of both places are tickled to see their towns mentioned. In fact, only persons whose towns aren't shown are likely to complain.)

Your second client, Karen Torricelli, is too far out of town even to think about convenient accessibility. Her business offers not a quick shopping trip but a recreational retreat: people go to Karen's Bowling Camp to get away from it all, to commune with nature, to hike and swim and fish, and to bowl. Although the map that meets Karen's needs might also benefit from the geometric distortion used on Rudy's map, it requires a different kind of geographic name-dropping.

Your map (fig. 5.7) offers an attractive solution for the Bowling Camp's brochure. Its prominent position at the top of the map fits with the local sense of its being "up north." You tout its accessibility via "scenic" Interstate 32 and imply that it is an attractive vacation spot for happy campers from as far away as Cleveland, Pittsburgh, and Knoxville. And you reproduce a miniature image of the highway exit sign to reinforce Karen's association with Lakeport and the Lake Walleye resorts. Of course, the only other resort you show is the comparatively upscale Kelly's Yacht Club, which also serves as a convenient landmark. Moreover, the town of Lakeport is "historic Lakeport." You give the area a cheery image with tiny pictorial symbols to mark an abandoned nineteenth-century military base and to suggest pleasant times to be had fishing in and sailing on Lake Walleye. Other tourist spots mentioned include the boyhood home of sage and renowned debtor Owen Moore, the unique, world-famous Stain Museum, and that delightful

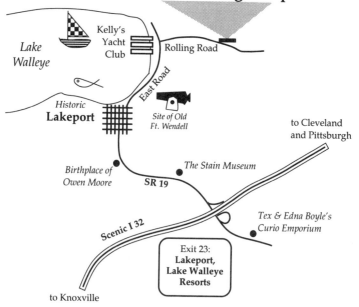

Karen's Bowling Camp

Lake Walleye

Kelly's Yacht Club

Rolling Road

East Road

Historic **Lakeport**

Site of Old Ft. Wendell

to Cleveland and Pittsburgh

Birthplace of Owen Moore

The Stain Museum

SR 19

Scenic I 32

Tex & Edna Boyle's Curio Emporium

Exit 23: **Lakeport, Lake Walleye Resorts**

to Knoxville

FIGURE 5.7. Map giving directions to Karen's Bowling Camp also shows the viewer other attractions in the area.

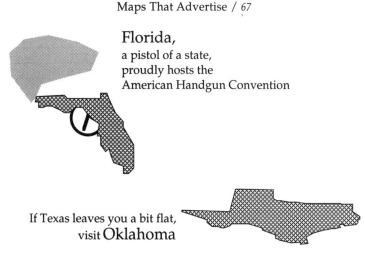

Florida,
a pistol of a state,
proudly hosts the
American Handgun Convention

If Texas leaves you a bit flat,
visit Oklahoma

FIGURE 5.8. Cartographic puns promoting a handgun convention in Florida (above) and in-state tourism among Oklahomans (below).

monument to Plastic America, Tex and Edna Boyle's Curio Emporium. Not exactly Lake Tahoe, central New Hampshire, or the Wisconsin Dells, the Lake Walleye area has a charm perhaps best appreciated by Tex and Edna's customers and Karen's guests. And your map captures this ambiance to perfection.

Spurred by the successful use of maps in your advertising campaigns for Rudy and Karen, you decide to reach out to potential clients elsewhere in the nation. Having seen a number of catchy ads using humorously distorted maps, you decide to make the cartographic pun a hallmark of your agency. After a few abortive attempts to capitalize on Scranton, Pennsylvania's, hourglass shape ("For the time of your life visit Scranton!") and the Miami–Fort Lauderdale, Florida, area's resemblance to an alligator ("Snap up your lot or condo today!"), you decide to work only with geographic shapes familiar to large numbers of nonlocal people. So after much thought and anguish you develop the kernels of two ideas, one for the Florida Convention Bureau and the other for the Oklahoma Convention Bureau. As figure 5.8 shows, the former is for a somewhat targeted—er, highly specific—campaign, which would not run in the general media, and the other is designed for obvious reasons to run only in Oklahoma, not in Texas. Hmmm. What could you do with New Jersey?

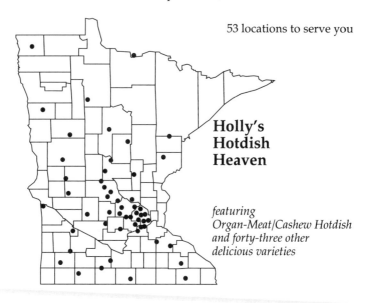

53 locations to serve you

**Holly's
Hotdish
Heaven**

*featuring
Organ-Meat/Cashew Hotdish
and forty-three other
delicious varieties*

FIGURE 5.9. Ad map proclaiming the widespread local acceptance of a regional restaurant chain.

Numerousness and Territoriality: Ad Maps Touting Success and Convenience

Not all your out-of-town clients have been as unappreciative as the Scranton and Miami tourist bureaus. Indeed, one of your biggest recent successes was a map-supported campaign developed for Holly's Hotdish Heaven, a regional restaurant chain with headquarters in Minneapolis. Holly's specializes in a Minnesota speciality, hotdish. In fact, all fifty-three restaurants are in Minnesota. (Unsuccessful attempts to establish outlets in Iowa, South Dakota, and Wisconsin suggest that the preference, or tolerance, for hotdish is highly regional.)

Anyway, your map (fig. 5.9) works for Holly's, which plans ten new outlets over the next five years, all in Minnesota. The map exploits a simple strategy common to maps advertising firms with multiple outlets or many far-flung clients: numerousness indicates success, and success indicates a superior product. The hungry Minnesotan has only to look at the map to see that Holly's has been successful. The dots indicate acceptance by discriminating palates throughout the state, but especially in the Twin Cities area toward the southeast, where more of the people live. Indeed, closing the three experimental restaurants in neighboring states meant that this single-state base map not only can show a fuller, more successful-looking packing of the dots but can also promote a stronger identification with the state and its people's pride in their traditions. If Holly's becomes much more successful, you'll have to use a smaller dot.

Another recent cartographic success is your map for Stanley Klutz Associates, a software developer and vendor active in geographic information systems (GIS). One of Klutz's best-selling products is a GIS designed to help states reapportion their congressional and legislative districts. Klutz redistricting software is widely recognized as providing an acceptable compromise that somehow meets the needs of powerful politicians now in office and satisfies the Supreme Court's notorious one person/one vote decision.

Eager to increase its market penetration, Klutz called after hearing of your spectacular success in the Minnesota hotdish campaign. Its marketing manager was particularly impressed by your map (fig. 5.10), which most effectively exploits Klutz's

States Using Stanley Klutz's Congressional and Legislative Redistricting System

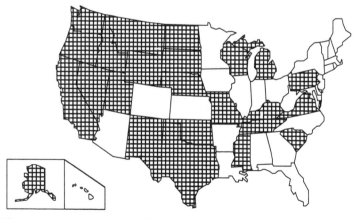

Most state governments are sticklers for StKCLRS.

FIGURE 5.10. Ad map equating numerousness and land area with product quality exploits sales to larger states in the West.

success among larger but often sparsely inhabited western states (some of which have only one congressman anyway). Indeed, your map creates the impression that well over half the country is using the Klutz system, when in fact twenty-six of fifty states use another vendor's system or the time-honored tradition of paper map, pencil, eraser, and the smoke-filled back room. Of course, if most of their success had been in the East, a map could have been linked to the slogan "Get with it, America—there's still lots of room for a Klutz!"

Like other artwork in commercial advertising, maps can be clever and catchy as well as contrived and deceptive. In most cases, though, the consumer recognizes the map as a playful put-on and appreciates being in on the joke. The ad map thus is further evidence of the map's enormous flexibility and appeal. Less benign, of course, are maps hawking remote building lots and doubtful mineral rights as well as many other maps that attempt to advocate or seduce. The next two chapters treat the use of maps as tools for real-estate developers and political propagandists.

Chapter 6

DEVELOPMENT MAPS
(OR, HOW TO SEDUCE THE TOWN BOARD)

Without maps, urban and regional planning would be chaotic. Detailed maps describe the relative size, shape, and spacing of a plan's components and suggest how well they interrelate. Maps of a planned shopping mall, for instance, would show the overall shape of the building, the sizes and general layouts of individual stores and public spaces, the size and locations of parking areas, and entrances to the parking lots and the mall building. Public officials use these maps to assess the impact of the proposed mall on nearby neighborhoods, traffic, and established businesses. Overlaying the mall plans onto topographic and soils maps reveals the mall's likely effects on wildlife, wetlands, and streamflow, as well as potential difficulties with sinkholes, unstable soils, or a high water table.

Even with maps, many cynics would argue, urban and regional planning is chaotic. As an inherently selective view of reality, the map often becomes a weapon in adversarial negotiations between developers and the local planning board. After all, the developer might have millions of dollars invested, whereas nearby residents probably don't want a new mall in their backyards, and planning boards often reflect local fears and biases. The developer's maps thus attempt to impress residents of more distant neighborhoods with the mall's elegance and convenience and to demonstrate that it will harm neither wildlife nor property values. If they use them at all, the residents' maps will focus on habitat destruction, traffic congestion, visual blight, noise, and trash. Because the developer has deeper pockets, the antimall maps will not look as nice as the promall maps, which suppress dumpsters, litter, and abandoned cars but optimistically portray skinny saplings as mature shade trees.

This chapter examines the role of maps in city and regional planning, especially as a tool of persuasion. It begins with a concise introduction to the part mapping plays in zoning and environmental protection, examines how developers might manipulate maps to enhance their cases, and concludes with a short example of how an overtaxed homeowner can use maps to argue for a reduced real-property assessment.

Zoning, Environmental Protection, and Maps

Zoning refers to the legal process municipalities use to control land use and land subdivision for development. Although legislation controlling the operation and powers of local planning and zoning boards varies from state to state, in general zoning laws allow regulation of the height, size, character, and function of buildings; the minimum size of a building lot and the location of the building on the lot; and associated improvements such as outbuildings, driveways, and parking lots. This system of laws, hearings, and enforcement procedures, which gained wide acceptance early in the twentieth century, is essential for effective urban planning.

Community planning boards commonly work with three principal maps: (1) an *official map* to show existing rights-of-way, administrative boundaries, parks and other public lands, and drainage systems; (2) a *master plan* to indicate how the area should look after several decades of orderly development; and (3) a *zoning map* to show current restrictions on land use. Figure 6.1 (pl. 4) shows the legend and a portion of a typical zoning map for a town far enough from a major city to still have working farms. The various categories reflect differences in the density of people and dwelling units, whether the area is generally open to the public and offers a pleasant view, and the likelihood that noise, litter, or other nuisances might affect adjoining areas. For each zoning category listed in the key, a set of restrictions applies to define precisely what kinds of structures and activities are permitted. In an R-1 residential district, for instance, a lot must have a minimum size of 5,000 square feet and a minimum width of 45 feet and may contain a single-family home that covers no more than 50 percent of

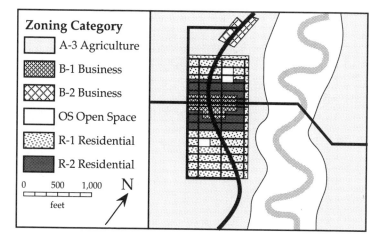

Zoning Category

- A-3 Agriculture
- B-1 Business
- B-2 Business
- OS Open Space
- R-1 Residential
- R-2 Residential

0   500   1,000   N

feet

FIGURE 6.1. Portion of a zoning map of a small community in a largely rural area.

the lot and is set back at least 20 feet from the front of the lot and 5 feet from each side.

Zoning boards spend considerable time hearing requests for *variances* so that property owners can do things that otherwise would violate the zoning ordinance. For example, a homeowner might want to add a room or a garage that would encroach within 10 feet of the property line. Or a business might want to purchase and raze a neighboring structure, grade the land, and make a parking lot. Zoning laws recognize that any improvement affects far more than just the ground underneath, but these laws also allow exceptions that meet community approval and are not too obnoxious. The zoning board must examine plans describing proposed alterations, then solicit and listen to the opinions of neighbors and other interested citizens. A board can grant or deny variances as proposed, grant modified variances with specific restrictions added, or grant temporary variances. The applicants or their neighbors can appeal the board's decision to a board of zoning appeals or to the court.

Major modifications of the zoning map are treated as modifications of the master plan and require a hearing before the planning board. Housing developments, new subdivisions, shopping malls, large stores, and other types of development that add or alter streets or affect municipal services such as water and sewage require planning board approval. The developer of a shopping mall must demonstrate, for instance, that runoff from the mall's parking lots will not cause flooding and that existing roads can carry the increased traffic. Sometimes the developer must work with the county or state highway department in planning new access roads, to be paid for directly or through an "impact tax." The developer of a subdivision must provide a map describing lot boundaries and showing roads and utility lines.

Planning boards are very concerned with housing density, that is, the number of dwellings per acre. Many municipalities specify a minimum lot size as large as three or five acres, ostensibly to "preserve the environment," but often to exclude low-income families unable to afford such expensive housing. But sometimes a planning board will approve a development of town houses or clusters of four or more dwellings with sufficient open space to meet a minimum *average* lot-size requirement. Maps are a convenient format for describing the developer's proposal and are an essential part of any subdivision hearing.

Growing concern about environmental degradation in the 1960s led to a substantial increase in environmental regulations and in the number and size of public agencies that monitor compliance. Although practices vary from place to place, state or municipal environmental quality review boards prepare inventories of vegetation, wetlands and other sensitive wildlife habitats, surface water, groundwater, soils, slope, geology, and historic sites. Commonly compiled using soil survey maps, aerial photography, or existing topographic maps, these *environmental resource inventories* are used to assess the likely adverse effects of highways, shopping malls, residential tracts, landfills, industrial plants, and other types of development. If the inventory is accurate, town or county planners can quickly tell whether a proposed project is likely to affect a fragile wetland habitat or contribute significantly to the extinction of a rare plant.

Large projects, whether public or private, commonly require an *environmental impact statement* (EIS). The developer usually must supplement information in the environmental resource inventory with field measurements compiled by an environ-

mental consulting firm of civil engineers, landscape archi-
tects, geologists, and biologists. The list of possible impacts
the EIS might address includes air pollution, water pollution,
public health, hydrology, erosion, geologic hazards such as
earthquakes and landslides, wildlife in general, flora (especially
rare plants), fauna (especially endangered species), aesthetic
and scenic values of both the natural landscape and the built
environment, solid waste, noise, the social environment, economic conditions, recreation, public utilities, transportation,
and the risk of accidents. An EIS must identify the types and
severity of plausible impacts, the areas affected, and alternative
strategies with a lesser impact.

Because preparing an EIS is costly and time-consuming, in
many cases a shorter, less comprehensive *environmental assessment* can demonstrate that the impact will be minimal, that the
project will comply with environmental regulations, and that
an EIS is not necessary. Environmental resource inventories
can enable local or state environmental review boards to evaluate
the environmental assessment and either approve the project
or require a full EIS.

Maps are an important part of an EIS or environmental
assessment. Environmental scientists commonly transfer all
map information to a common base for ready comparison.
Sometimes a computerized geographic information system
(GIS) is used to store the data and generate final plots. Detailed,
oversize maps might accompany the EIS in an appendix, to
supplement smaller-scale, more generalized maps in the body
of the report. (How many readers bother to compare the
large-scale maps with their small-scale generalizations, positioned much closer to the analytical and persuasive parts of
the report?) Potentially significant sources of error are the
transfer of information from the source map to the common
base and the generalization of these small-scale maps. (When
the alternatives are equally valid, can the consultant resist the
temptation to draw the line that favors the client's case?)
Additional problems arise when boundaries and other data
are transferred from unrectified aerial photographs (see chap.
3)—in hilly areas these lines are probably not accurate unless
the mapmaker used a stereoplotter. Persons opposing a project
might well begin by taking the EIS into the field and checking
its supporting maps against the features portrayed.

Sometimes the data simply cannot reflect what the developer or the compliance agency would like them to show. As examples, floodplains defined locally by a single elevation contour tend to include either too much or too little of the real floodplain, and soil survey maps might not reliably reflect the depth to bedrock, an important indicator of where septic tanks and leach fields are unsuitable. When using soils maps to compile maps for an environmental assessment, the developer should resist the temptation, illustrated in figure 6.2, to alter categories, replace a technical definition with a questionable interpretation, or "inadvertently" omit small parcels of land whose presence might disqualify a larger tract. Both the developer and the review panel should be aware that soils maps are based on a soil scientist's interpretation of the land surface and a limited number of subsurface core samples. Moreover, soils maps generally do not show patches smaller than the head of a pencil, which at 1:20,000 would cover several generous building lots. When the developer's consulting engineer carries out a special field survey, a map should show the actual locations of the subsurface core samples to allow the viewer to verify that the map is neither much more nor much less generalized than the data it represents.

Zoning cases and environmental quality reviews often move from the administrative hearing to the courtroom. When deliberations escalate to the judicial level, maps become important as exhibits, together with aerial photographs, drawings

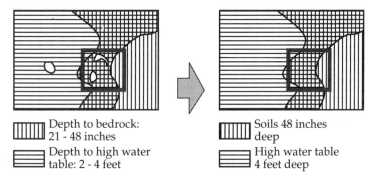

FIGURE 6.2. Creative generalization and interpretation of the soils map and categories on the left can yield the more favorable cartographic representation on the right.

and architect's renderings, scale models, ground-level photographs, movies, and videotapes. The courtroom audience usually is more sophisticated and less harried than the local planning or zoning board, and both parties commonly employ technical experts to testify as well as to advise in cross-examination. Ploys that might have impressed a volunteer group of business people, farmers, teachers, and homemakers can fail or backfire, and exhibits must be easily defended as well as convincing and persuasive. Opposing sides often differ principally in their interpretation of identical exhibits.

Maps are indispensable exhibits in land-use cases, which cannot be prepared and presented without them. A minimal presentation would include the master plan, a detailed zoning map of the affected area, and an enlarged aerial photograph or two. Attorneys and witnesses often use a marking pen or tape to identify locations, and they require multiple copies of most exhibits so that these marked-up materials can become part of the record. The attorney carefully marks each exhibit for introduction as evidence, identifies its source, calls an appropriate witness to explain it, and reviews in advance the scope of the witness's testimony and plausible counterarguments by the opponent.

Although topographic maps, air photos, and zoning maps tolerate no embellishment, the developer has considerable license in the design and content of site plans and architect's renderings. A particularly interesting and forceful graphic is the *concept diagram*, a schematic, somewhat stylized map intended to demonstrate the general layout and functional relationship of a plan's main elements. The example in figure 6.3, which illustrates the interchange concept for a downtown transportation center, shows how the developer or planner uses lines to subdivide space, highlight patterns of movement, and suggest revitalization of the central city. Concept diagrams have a compelling, mysterious attraction and can be highly persuasive when explained by an enthusiastic architect. These maps encourage the viewer to want to see the plan work, not to wonder whether it will work. Once the concept diagram has convinced the audience that a project is functional and feasible, the presenter can introduce three-dimensional models, sketches, and other persuasive renderings to show how the finished project should look. Like most utopian

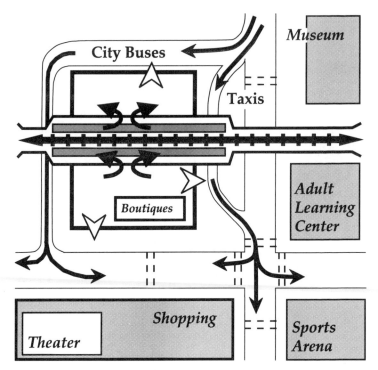

FIGURE 6.3. A concept diagram for a proposed downtown transportation center with rail, bus, and taxi service.

views of the future, these maps and pictures feign realism with selective detail.

Eleven Rules for Polishing the Cartographic Image

With the following eleven rules the developer can play down the adverse impact of a proposed project and enhance its visual appearance and presumed benefits. These guidelines work best before a town zoning or planning board and are least effective in state or federal court. (It would be a shame, after all, for the truly cynical land developer or waste-treatment consultant not to take advantage of the public's graphic naïveté and appalling ignorance of maps.)

1. *Be shrewdly selective.* Don't show what you'd rather they not see. Omit potentially embarrassing features that

might evoke unpleasant images of litter, congestion, and noise. Omit dumpsters and other trash containers, traffic signs, loitering teenagers, and trucks. In sketches and scale models either omit people and cars altogether or show only smartly dressed people and late-model cars and station wagons. Never admit the possibility of dying shrubbery, trampled turf, or anything remotely suggesting a plume of smoke. Above all, keep the image clean and sufficiently generalized so that such omissions don't appear unnatural.

2. *Frame strategically.* Avoid unfavorable juxtaposition, and crop the maps and sketches to forestall fears of illness or diminished property values. If a neighboring site is unattractive or likely to be unfavorably affected, leave it out. If the proposed development adjoins a park or another attractive site, leave it in. If a neighboring property might also be improved, include it but show it too as newly developed. Never show the school or homes bordering a proposed landfill, solid waste treatment facility (the new term for municipal incinerator), or brewery.

3. *Accentuate the positive.* Choose favorable data and supportive themes for maps. If, for instance, a proposed landfill will have a high fence or unobtrusive entrance, by all means show it. If a new mall would displace an existing eyesore, a set of "before" and "after" maps is useful. Favorable interpretations of data or source maps also help.

4. *If caught, have a story ready.* Computer errors, a stupid drafting technician's use of the wrong labels, or the accidental substitution of an earlier version of the map make plausible excuses.

5. *Minimize the negative.* If you can't eliminate them entirely, at least don't emphasize features you'd rather have ignored. Note that the train station in figure 6.3 doesn't call attention to exhaust from idling buses and taxis.

6. *Dazzle with detail.* After all, a detailed map is a technically accurate map, right? Details are useful distractions.

7. *Persuade with pap.* Try highly simplistic maps, or maps with fire hydrants, mailboxes, and any other irrelevant minutiae that might camouflage potentially embarrassing details.

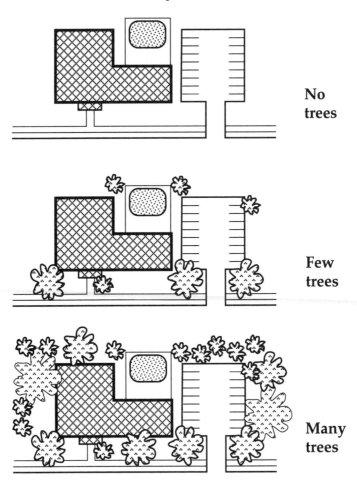

FIGURE 6.4. Tree symbols add visual appeal to an otherwise barren developer's plan.

8. *Distract with aerial photographs and historical maps.* These make great conversation pieces and are excellent distractions for people eager to exclaim, "Hey, there's my house!"

9. *Generalize creatively.* Filter or enhance details to prove your point. A little selective omission or massaging of contours, soils boundaries, or even property lines might well pass for cartographic license.

10. *Enchant with elegance.* And don't forget the architect's cartographic friend, the tree stamp. As figure 6.4 dem-

onstrates, symbolic trees can convert a mundane proposal into a pleasant neighborhood asset, and the more of these hypothetical trees, the better. After all, it takes much less time and effort to stamp or paste treelike symbols onto the map than to plant the real thing. And in twenty years those anemic saplings you will plant might even resemble the healthy shade trees in the picture.

11.*When all else fails, try bribery.* Not under-the-table monetary payoffs, of course, but such institutional bribery as decent-paying jobs for the unemployed, good profits and well-paying jobs for contractors and construction workers, a larger tax base or "payments in lieu of taxes" for local government, prestige, and promises of amenities elsewhere in the community for teenagers, young families, and senior citizens. Or try another area, where citizens and their representatives are less aware of graphic trickery.

Your Turn:
The Assessment Review

Like words and numbers, maps are anybody's weapon, and they can also help the homeowner appeal an unfairly high tax assessment. But whether the advice that follows is useful will depend upon how the area where you buy estimates real-estate values.

In most areas, the municipality computes the yearly tax on "real property" by multiplying the assessed value of a building and lot by the established local tax rate. The total value of land and buildings in the area and its estimated expenses for schools, government operations, welfare, and debt service determine the tax rate. Each parcel's assessed value is someone's guess of what the property is worth. In areas with "full-value assessment," this guess is an estimate of the current fair market value of the property. In other areas the assessed value is supposed to be a fixed percentage of the fair market value. But other factors, some political, often influence the guessing.

Assessment practices vary widely. Some jurisdictions rigorously apply a set of guidelines, or formulas, based on measurements such as lot size, area of the dwelling, and number of bedrooms and bathrooms. Other jurisdictions filter these objective criteria through the subjective judgment of an assessor or assessment board. Since location is an important part of a property's worth, assessors not slavishly governed by formulas will consider scenic values and other less tangible amenities as well as recent sale prices of nearby properties.

Because housing values change over time in response to inflation, new amenities and nuisances, and shifting perceptions of which neighborhoods are good and which aren't, assessors usually spend more time reassessing old properties than evaluating new ones. Some areas reassess yearly or on a regular cycle of every two, three, or four years, whereas others reassess more randomly, according to the judgment of the assessor.

A common practice is to reassess only when a property is sold or altered by a major structural change such as adding a room or a fireplace. In these areas assessments tend to favor long-term residents who bought property many years ago and have made no major improvements. Widespread reassessment would tend to hurt these people, especially if their neighborhoods have not deteriorated. Because long-term residents tend to oppose new assessment practices that might be fairer and also to vote regularly, a directly elected assessor or one appointed by the town board is likely to reflect their interests. The result is an assessment scheme aptly called "soak the newcomer" or "Welcome, stranger."

Fear not, though, for assessments, like zoning decisions, can be appealed. In larger municipalities, a formal appeal might even require the services of an attorney. Towns and villages, more likely to reassess mostly newcomers, have a more relaxed approach—an annual "grievance day" that might even be a week or a month long. Residents come before the board to "grieve," that is, to challenge their assessments. Yet a successful appeal will depend not on griping about the assessor's incompetence or the cost and quality of municipal services but on demonstrating convincingly that the proposed assessment clearly is out of line. In short, you need to do some research.

The assessor's office is a good place to begin. To show that you have been overassessed, you might compare your assessment with those of other properties in the neighborhood, especially similar houses. Come prepared with the addresses of properties whose records you want to examine. Workers in the assessor's office will help you locate a description of each property and its assessment history. These records show note-

Homes Similar to 121 Millard Fillmore Drive

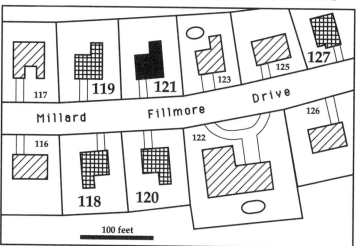

FIGURE 6.5. A map showing your home and nearby houses used for comparison.

worthy improvements and special features and list the sizes of the lot, the living area, and perhaps individual rooms. If your assessment is not radically different from those of similar dwellings, you probably have no case—but count yourself fortunate to have been treated fairly from the outset. Yet if your assessment is well above those of similar properties nearby, an appeal could save you considerable money, especially if you stay for several years.

Present your appeal using two types of evidence: facts showing that the properties you think are similar are indeed so, and comparative figures demonstrating your new assessment is too high. (Should you find that one of the similar properties is also taxed unfairly, you might ask your neighbor for support in a joint appeal.) Three kinds of exhibits can help establish similarity: a map of the neighborhood, a poster with photographs and street addresses, and tables or charts comparing your house with the others for type of construction (wood frame, brick, stone, . . .), number of rooms, area of living space, year of construction, and lot size. Figure 6.5, a typical neighborhood map compiled by tracing street, lot, and foundation lines from the assessor's maps, can show both similarity and proximity. If presented in a large poster, this

Assessment Disparities on Millard Fillmore Drive

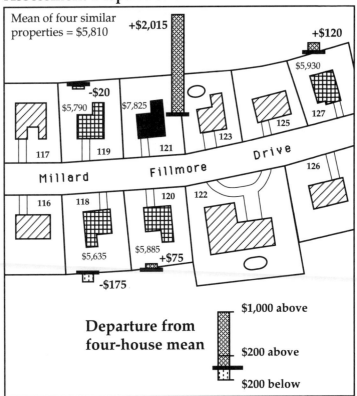

FIGURE 6.6. A map showing disparities in assessed value between your house and similar properties in the neighborhood.

map makes a good prop for discussing other data that you provide in a table, with a copy for each board member. As an able propagandist, you naturally choose a title such as "Similar Properties in the Neighborhood" to reinforce your point.

Having convinced the board that these houses are similar to your own, you now present a map dramatizing disparities in assessment. As in figure 6.6, the mean assessment for the group of similar properties (excluding yours) can be a useful basis for graphic comparison as well as a subtle hint of a fair assessment for your own parcel. As a referent, this mean assessment also allows you to use symbols that focus on *differences*

Is 121 really the most valuable property on Millard Fillmore Drive?

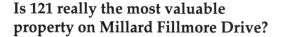

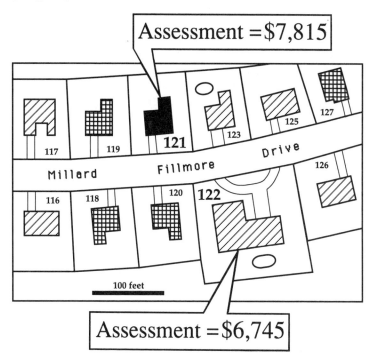

FIGURE 6.7. A map to demonstrate that your proposed assessment ought not be the highest in the neighborhood.

in assessments, so that comparative bar symbols more effectively represent unfairness. If you can win the argument for similarity, this map can win your case for a lower assessment.

If your proposed assessment exceeds that of any of your neighbors, you might want to lead by comparing your property with the parcel having the next highest assessment. For this presentation use a simple map showing lot locations, addresses, and assessments. As figure 6.7 illustrates, the map's symbols and labels should show board members that you know about assessments in your neighborhood and that you have a legitimate grievance. Then introduce photographs and tables showing that your house is much more modest in appearance, size, and amenities than the one with the second highest assessment. If the assessor has flagrantly been playing "soak the newcomer," your exhibits will send a clear message that you are fully prepared to embarrass the board unless they make an appropriate adjustment. If you use tact, most of time they will.

Although the tone of this chapter is cynical, the intent is to make you skeptical about how some people use maps, not cynical about maps in general. Understanding cartographic manipulation is important to being an informed citizen able to evaluate a wide range of proposals for altering the landscape and the environment. In viewing maps it is essential to remember that a particular view of reality (or a future reality) is not the only view and is not necessarily a good approximation of truth.

Chapter 7

MAPS FOR POLITICAL PROPAGANDA

A good propagandist knows how to shape opinion by manipulating maps. Political persuasion often concerns territorial claims, nationalities, national pride, borders, strategic
positions, conquests, attacks, troop movements, defenses,
spheres of influence, regional inequality, and other geographic
phenomena conveniently portrayed cartographically. The
propagandist molds the map's message by emphasizing
supporting features, suppressing contradictory information,
and choosing provocative, dramatic symbols. People trust
maps, and intriguing maps attract the eye as well as connote
authority. Naive citizens willingly accept as truth maps based
on a biased and sometimes fraudulent selection of facts.

Although all three manipulate opinion, the propagandist's
goals differ from those of the advertiser and the real-estate
developer. Both the advertiser and the political propagandist
attempt to generate demand, but the advertiser sells a product
or service, not an ideology. Both the advertiser and the propagandist attempt to lower public resistance or to improve a
vague or tarnished image, but the advertiser's objectives are
commercial and financial, whereas the propagandist's are
diplomatic and military. Both the real-estate developer and
the political propagandist seek approval or permission, but
the developer is concerned with a much smaller territory,
often uninhabited, and seldom acts unilaterally without official
sanction. Although both the real-estate developer and the
propagandist face opponents, the developer usually confronts
groups of neighboring property owners, environmentalists, or
historic preservationists, whereas the propagandist commonly confronts a vocal ethnic minority, another country, an alliance
of countries, an opposing ideology, or a widely accepted

standard of right and wrong. Because propaganda maps are more likely to be global or continental rather than local, the political propagandist has a greater opportunity than either the advertiser or the real-estate developer to distort reality by manipulating the projection and framing of the map.

This chapter explores the map's varied and versatile role as an instrument of political propaganda. Its first section examines how maps function as political icons—symbols of power, authority, and national unity. Next the chapter looks at how map projections can inflate or diminish the area and relative importance of countries and regions, and how a map projection can itself become a rallying point for cartographically oppressed regions. A third section examines the manipulations of Nazi propagandists, who used maps to justify German expansion before World War II and to try to keep America neutral. A final section focuses on a few favorite symbols of the cartographic propagandist: the arrow, the bomb, the circle, and place-names.

Cartographic Icons Big and Small: Maps as Symbols of Power and Nationhood

The map is the perfect symbol of the state. If your grand duchy or tribal area seems tired, run-down, and frayed at the edges, simply take a sheet of paper, plot some cities, roads, and physical features, draw a heavy, distinct boundary around as much territory as you dare claim, color it in, add a name— perhaps reinforced with the impressive prefix "Republic of"— and presto: you are now the leader of a new sovereign, autonomous country. Should anyone doubt it, merely point to the map. Not only is your new state on paper, it's on a map, so it must be real.

If this map-as-symbol-of-the-state concept seems farfetched, consider the national atlases England and France produced in the late sixteenth century. Elizabeth I of England commissioned Christopher Saxton to carry out a countrywide topographic survey of England and Wales and to publish the maps in an elaborate hand-colored atlas. In addition to providing information useful for governing her kingdom, the atlas bound together maps of the various English counties and asserted

FIGURE 7.1. Engravings reflect the iconic significance of maps and atlases as national symbols in Christopher Saxton's 1579 *Atlas of England and Wales* (left) and Maurice Bouguereau's 1594 *Le théâtre françoys* (right).

their unity under Elizabeth's rule. Rich in symbolism, the atlas's frontispiece (fig. 7.1, left) was a heavily decorated engraving that identified the queen as a patron of geography and astronomy. A few decades later, Henry IV of France celebrated the recent reunification of his kingdom by commissioning bookseller Maurice Bouguereau to prepare a similarly detailed and decorated atlas. Like Saxton's atlas, *Le théâtre françoys* includes an impressive engraving (fig. 7.1, right) proclaiming the glory of king and kingdom. In both atlases regional maps provided geographic detail and a single overview map of the entire country asserted national unity.

The spate of newly independent states formed after World War II revived the national atlas as a symbol of nationhood. Although a few countries in western Europe and North America had state-sponsored national atlases in the late nineteenth and early twentieth centuries, these served largely as reference works and symbols of scientific achievement. But between 1940 and 1980 the number of national atlases increased from fewer than twenty to more than eighty, as former colonies turned to cartography as a tool of both economic development and political identity. In the service of the state, maps and atlases often play dual roles.

Perhaps the haste of new nations to assert their independence cartographically reflects the colonial powers' use of the map as an intellectual tool for legitimizing territorial conquest, economic exploitation, and cultural imperialism. Maps made it easy for European states to carve up Africa and other heathen lands, to lay claim to land and resources, and to ignore existing social and political structures. Knowledge is power, and crude explorers' maps made possible treaties between nations with conflicting claims. That maps drawn up by diplomats and generals became a political reality lends an unintended irony to the aphorism that the pen is mightier than the sword.

Nowhere is the map more a national symbol and an intellectual weapon than in disputes over territory. When nation A and nation B both claim territory C, they usually are at war cartographically as well. Nation A, which defeated nation B several decades ago and now holds territory C, has incorporated C into A on its maps. If A's maps identify C at all, they tend to mention it only when they label other provinces or subregions. If nation B was badly beaten, its maps might show C as a disputed territory. Unlike A's maps, B's maps always name C. If B feels better prepared for battle or believes internal turmoil has weakened A, B's maps might more boldly deny political reality by graphically annexing C.

Neutral countries tread a thin cartographic line by coloring or shading the disputed area to reflect A's occupation and perhaps including in smaller type a note recognizing B's claim. If A and B have different names for C, A's name appears, sometimes with B's name in parentheses. (Even when recapture by B is improbable, mapmakers like to hedge their bets.) Cartographic neutrality can be difficult, though, for customs officials of nation B sometimes embargo publications that accept as unquestioned A's sovereignty over C. If A's rule is secure, its censors can be more tolerant.

Consider, for example, the disputed state of Jammu and Kashmir, lying between India, Pakistan, and China. Both India and Pakistan claimed Kashmir, once a separate monarchy, and went to war over the area in August 1965. Figure 7.2, a U.S. State Department map, shows the cease-fire line of fall

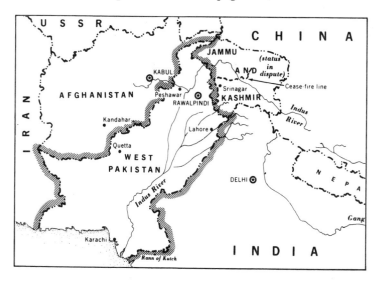

FIGURE 7.2. Disputed India-Pakistan boundary and the territory of Jammu and Kashmir, as portrayed in the 1965 *Area Handbook for Pakistan*, published by the U.S. government.

1965, which placed Pakistan in control of northwestern Kashmir and showed India in control of the southern portion. (China occupied a portion of northeastern Kashmir.) Nonetheless, Indian and Pakistani maps continued to deny political reality. A 1984 Pakistani government tourist map (fig. 7.3, lower), for instance, included Kashmir in Pakistan, whereas a map (fig. 7.3, upper) in an Indian government tourist brochure ceded the entire territory to India. American and British atlases attempted to resolve the dispute with notes identifying the area occupied by Pakistan and claimed by India, the area occupied by India and claimed by Pakistan, three areas occupied by China and claimed by both India and Pakistan, the area occupied by China and claimed by India, and the area occupied by India and claimed by China. And for years publishers found it difficult to export the same books on South Asian geography to both India and Pakistan.

Even tiny maps on postage stamps can broadcast political propaganda. Useful both on domestic mail to keep aspirations alive and on international mail to suggest national unity and determination, postage stamp maps afford a small but numerous means for asserting territorial claims. As shown in

FIGURE 7.3. Official government tourist maps show Kashmir as a part of India (above) and as a part of Pakistan (below). In reality, India controls the southern part of the state of Kashmir, Pakistan controls the northwestern part, and China controls three sections along the eastern margin.

FIGURE 7.4. Subtle and not-so-subtle cartographic propaganda on Argentinian postage stamps.

figure 7.4, Argentinian postage stamps have touted that nation's claims not only to the Falkland Islands and the British-held islands to their east but also to Antarctica. Like all official maps of Argentina, these postage stamps deny the legitimacy of British occupation with their Spanish label "Islas Malvinas." Postage stamps bearing maps are also useful propaganda tools for emergent nations and ambitious revolutionary movements.

Size, Sympathy, Threats, and Importance

Sometimes propaganda maps try to make a country or region look big and important, and sometimes they try to make it look small and threatened. In the former case, the map might support an appeal to fairness: the Third World is big, and therefore it deserves to consume a larger share of the world's resources, to exercise more control over international political bodies such as UNESCO (the United Nations Educational, Scientific, and Cultural Organization), and to receive greater respect and larger development grants from the more developed nations of the West and the Communist world. In the latter case, the map might dramatize the threat a large state or group of states poses for a smaller country. Figure 7.5, for instance, portrays a cartographic David-and-Goliath contest between tiny Israel and the massive territory of the nearby oil-rich Arab nations. Even though the map's geographic facts are accurate, a map comparing land area tells us nothing about Israel's advanced technology, keen military preparedness, and alliances with the United States and other Western powers.

Some map projections can help the propagandist by making small areas bigger and large areas bigger still. No projection has been as abused in the pursuit of size distortion as that devised by sixteenth-century atlas publisher and cartographer Gerardus Mercator. Designed specifically to aid navigators, the Mercator projection vastly enlarges poleward areas so that straight lines can serve as *loxodromes*, or *rhumb lines*—that is, lines of constant geographic direction. (If the navigator's compass shows true north rather than magnetic north, rhumb lines can be called lines of constant compass direction.) As figure 7.6 shows, the navigator finds the course by drawing a straight

FIGURE 7.5. Map showing the encirclement of Israel by neighboring Arab nations, redrawn from a map published during the 1973 war by the Jewish National Fund of Canada.

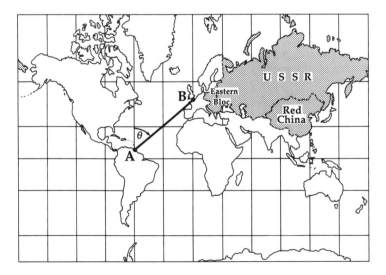

FIGURE 7.6. Mercator world map showing the bearing angle Θ for a rhumb line from A to B and the areal exaggeration of Red China and in particular the USSR. Designed to aid navigators, the Mercator also has served political propagandists seeking to magnify the Communist threat.

line from origin A to destination B and then reading the angle θ from the meridian to the rhumb line. If one consistently follows this bearing from A, one will eventually reach B. For this convenience the navigator must sacrifice a shorter but less easily followed great-circle route and endure the areal distortion caused by the progressive increase poleward of north-south scale. In fact, the projection shows little of the area within the Arctic Circle and the Antarctic Circle because its poles are infinitely far from its equator. Ever wary of icebergs anyway, navigators for centuries have avoided polar waters and accepted as only a minor liability the Mercator projection's gross areal exaggeration. Yet for decades the John Birch Society and other political groups intimidated by Communist ideology and Stalinist atrocities have reveled in the Mercator's cartographic enhancement of the Soviet Union. Birch Society lecturers warning of the Red menace commonly shared the stage with a massive Mercator map of the world with China and Russia printed in a provocative, symbolically rich red.

Although equal-area map projections (as in figs. 2.5 and 2.6) have been available at least since 1772, when Johann Heinrich Lambert published his classic *Beiträge zum Gebrauche der Mathematik und deren Anwendung*, Mercator's projection provided the geographic framework for wall maps of the world in many nineteenth- and early twentieth-century classrooms, and more recently for sets of television news programs and backdrops of official briefing rooms. Perhaps distracted by concerns with navigation, exploration, and time zones, cartographically myopic educators and set designers presented a distorted world view that diminished the significance of tropical areas to the advantage of not only Canada and Siberia but western Europe and the United States as well. The English especially liked the way the Mercator flattered the British Empire with a central meridian through Greenwich and prominent far-flung colonies in Australia, Canada, and South Africa. Some British maps even gave the Empire an added plug by repeating Australia and New Zealand at both the left and right sides of the map.

Yet in the early 1970s this subtle and probably unwitting geopolitical propaganda served as a convenient straw man for German historian Arno Peters, who published a "new" world map based on an equal-area projection similar to one described in 1855 by the Reverend James Gall, a Scottish clergy-man. As figure 7.7 shows, the Gall-Peters projection gives tropical continents a mildly attenuated, stretched look, which probably explains why geographers and cartographers have adopted more plausible equal-area maps and why the basic texts on map projections Peters consulted had ignored Gall's contribution. Indeed, Lambert and other cartographers had developed numerous equal-area map projections, including many that distorted shape much less severely than does the Gall-Peters version.

But Dr. Peters knew how to work the crowd. A journalist-historian with a doctoral dissertation on political propaganda, Peters held a press conference to condemn the Mercator world view (as well as all nonrectangular projections) and to tout his own projection's "fidelity of area" and more accurate, "more egalitarian" representation of the globe. By calling attention to the Mercator's slighted portrayal of most Third World nations and blaming a stagnation in the development of cartography, Peters struck responsive chords at the World Council of Churches, the Lutheran Church of America, and various United Nations bodies. Religious and international development organizations welcomed Peters and his "new cartography," with the greater fairness and accuracy it promised. They also pub-

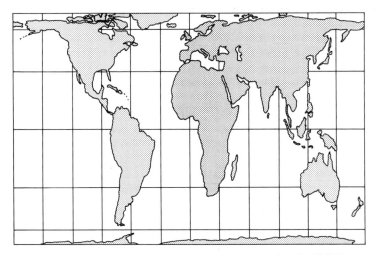

FIGURE 7.7. The Peters projection or, more accurately, the Gall-Peters projection.

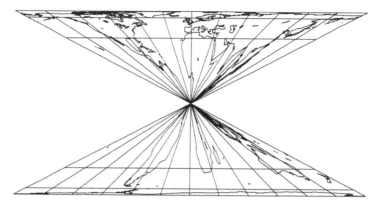

FIGURE 7.8. Like all equal-area projections, this hourglass equal-area map projection John Snyder devised as a joke has area fidelity but distorts shape.

lished large and small versions of the Peters projection, hung it on their walls, and used it in their press releases and publications. Perhaps because journalists also like to champion the oppressed and can't resist a good fight, the press repeated Peters's claims and reported the success of his bandwagon. Academic cartographers became both puzzled and enraged— puzzled that the media and such prominent, respected institutions could be so gullible and ignorant, and enraged that these groups not only so persistently repeated Peters's preposterous assertions but so obstinately refused to look at cartography's writings, accomplishments, and rich history.

Not all cartographers lacked a sense of humor. U.S. Geological Survey cartographic expert John Snyder, himself a developer of several useful as well as innovative map projections, offered yet another equal-area projection to underscore his cartographic colleagues' point that an equal-area map is not necessarily a good map. As shown in figure 7.8, Snyder's hourglass equal-area projection does what the Peters projection does and the Mercator doesn't—it preserves areal relationships. But it also demonstrates dramatically that areal fidelity does not mean shape fidelity.

Ironically, by succumbing to Peters's hype, UNESCO and other organizations sensitive to Third World problems loyally backed the wrong projection and missed an enormous propaganda opportunity. By accepting uncritically the rather dubious assumption that a map responsive to people should accurately represent land area, these groups not only demonstrated

a profound cartographic naïveté but also ignored a more hu-
manistic type of map projection that actually makes some
Third World populations appear justifiably enormous. How
much more convincing their media blitz might have been
had they supported a demographic base map, or area cartogram, similar to figure 2.10, on which the area of each country is scaled according to number of inhabitants. Indeed, an
area cartogram would be more effective than the Peters projection in boosting the importance of China, India, and Indo-
nesia and in revealing the less substantial populations of
Canada, the United States, the Soviet Union, and other comparatively less crowded countries. But perhaps a more subtle
internal need motivated leaders of UNESCO and the World
Council of Churches, for the Peters projection is comparatively kinder to the low and moderate population densities of
Africa, Latin America, and the Middle East—indeed, a cynic
might note the influence of African diplomats in UNESCO
and the inherent interest of the World Council of Churches in
concentrated Christian missionary activity in Latin America
and central Africa.

Propaganda Maps and History:
In Search of Explanation and Justification

Although propaganda cartography is probably not much
younger than the map itself, the Nazi ideologues who ruled
Germany from 1933 to 1945 warrant special mention. No
other group has exploited the map as an intellectual weapon
so blatantly, so intensely, so persistently, and with such variety. Nazi propaganda addressed especially to the United
States presented a selective and distorted version of history
designed to increase sympathy for Germany, decrease support
for Britain and France, and keep America out of World War
II, at least until Axis forces had conquered Europe. The
examples discussed in this section are from a weekly news
magazine, *Facts in Review*, published in New York City during
1939, 1940, and 1941 by the German Library of Information.
   The sympathy theme of Nazi cartopropaganda often re-
called Germany's defeat in World War I—a humiliation followed
by an economic depression that helped the National Socialists
to power. Figure 7.9, which compared the German plight in

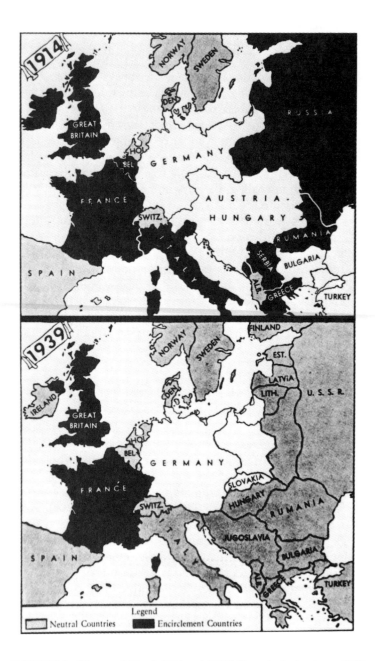

FIGURE 7.9. "Then and Now! 1914 and 1939" (*Facts in Review* 1, no. 17 [8 December 1939]: 1).

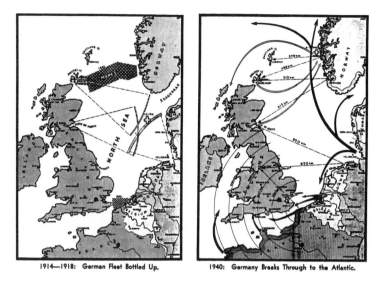

1914—1918: German Fleet Bottled Up.   1940: Germany Breaks Through to the Atlantic.

FIGURE 7.10. "The War in Maps" (*Facts in Review* 3, no. 16 [5 May 1941]: 250).

1914 with that of 1939, invoked a persistent anti-British theme. These two maps formed much of the front page of *Facts in Review* for 8 December 1939. A caption to the left of the 1914 map noted the encirclement that "provided the necessary basis for Britain's successful Hunger-blockade," whereas the caption for the 1939 map alluded to Britain's failed attempts to repeat the encirclement and proclaimed that "the path of industrial and economic cooperation to the East and the Southeast lies open!" Note, though, that the 1939 map conveniently groups Germany's main allies at the time, Mussolini's Italy and Stalin's Russia, with Switzerland and other "neutral countries."

In early 1941, another map attempted to explain and justify Germany's western advance against England into France, Belgium, and Holland by comparing Germany's strategic disadvantage in 1914 with the more favorable situation in 1940. Figure 7.10 contrasts the German navy "bottled up" by the British in the North Sea in 1914 through 1918 with the German navy that in 1940 had "[broken] through to the Atlantic." Hitler had not yet turned against Stalin, and the map's caption noted that whereas Germany had to fight on two, and later three, fronts in 1914, "Today no such danger exists. The British blockage is ineffective and, instead, the blockaders themselves are being blockaded." Arcs reinforce the blockade theme of the 1914–18 map, and bold arrows dramatize Germany's freer access to the Atlantic on the 1940 map.

Other Nazi maps attempted to divert sympathy from Britain. Captioned "A Study in Empires," the charts in figure 7.11 compare the 264,300 mi² on which Germany's 87 million inhabitants "must subsist" with the 13,320,854 mi² that Britain, with only 46 million people, "has acquired." How can little Germany be the aggressor nation? the left panel asks. In contrast, the right panel suggests a note of greed in Britain's conquest of 26 percent of the world's land area. The map's caption sounds a further chord of grievance by noting that the British Empire "includ[es] the former German colonies."

Facts in Review's editors also used maps to cast doubt on England's probity. In the issue of 30 November 1940, a story headlined "British Bombings—A Record of British Truthfulness" reported that on 24 November a British bomber apparently lost its way to Genoa and bombed Marseilles, France.

A STUDY IN EMPIRES

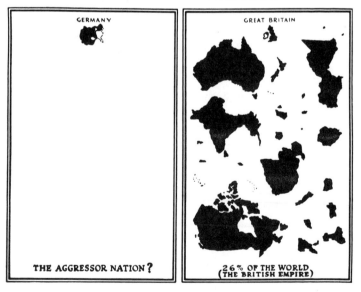

FIGURE 7.11. "A Study in Empires" (*Facts in Review* 2, no. 5 [5 February 1940]: 33).

FIGURE 7.12. "Marseille 'Mistaken' for Genoa" (*Facts in Review* 2, no. 46 [30 November 1940]: 566).

Early British news reports not only had denied the bombing but had blamed the Germans. A map (fig. 7.12) located both cities, and its caption reeked with sarcasm: "Marseille was 'mistaken' for Italy's Genoa, more than 200 miles away!" The story developed a bumbling-British theme by noting the dropping of anti-Italian leaflets, casualties of six dead and twelve wounded ("These 18 persons were exclusively women"), the protests of the Vichy government, and England's

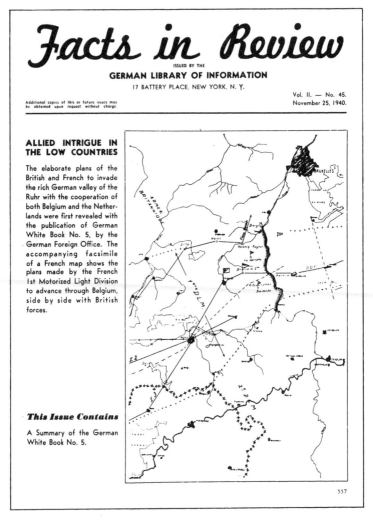

FIGURE 7.13. "Allied Intrigue in the Low Countries" (*Facts in Review* 2, no. 45 [25 November 1940]: 557).

"somewhat lame story that fog and inexperience caused the crew of the British plane to drop their bombs over this non-combatant city."

Nazi propagandists also used facsimile maps to prove their opponents' treachery and justify Germany's advancing western front. Nonskeptical Americans were thought likely to accept the largely illegible, hand-labeled map (fig. 7.13) on the *Facts in Review* cover for 25 November 1940 as convincing evidence of British and French plans to "invade the rich German valley of the Ruhr with the cooperation of both Belgium and the Netherlands." Germany, the map implied, had merely done to them first what they had been plotting to do to her.

Another plot revealed in *Facts in Review* justified the partition of Poland among Germany and Russia. Captioned "Polish Delusions of Grandeur," figure 7.14 shows in bold black a much reduced German state. Offended and outraged, the editors revealed that "this map, published in the Posen newspaper, 'Dziennik Poznanski,' after the receipt of Chamberlain's 'blank check,' revived dreams of extending the Polish dominion to the Weser River." Although a newspaper map hardly constitutes official state policy, the map suggests to the politically naive that the 1939 invasion amply repaid the Poles for even daring to think of annexing German territory.

Useful for representing one's opponents as the bad guys, maps can also advertise oneself as the good guy. Accompanying a story headlined "Repatriation: Background for Peace,"

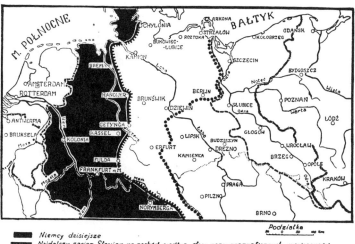

FIGURE 7.14. "Polish Delusions of Grandeur" (*Facts in Review* 2, no. 28 [8 July 1940]: 294).

FIGURE 7.15. "Repatriation: Background for Peace" (*Facts in Review* 1, no. 16 [30 November 1939]: 3).

figure 7.15 shows Germany the Peacemaker quietly reducing ethnic friction in the Baltic states by evacuating 80,000 to 120,000 Germans. As *Facts in Review* proudly observes, "Germany is not afraid to correct mistakes of geography and history." The map's pictorial symbols dramatize the repatriation by showing proud, brave, obedient Germans clutching their suitcases and lining up to board ships sent to "lead [these] lost Germans back home to the Reich." To the east in stark, depressing black looms the Soviet Union, and to the south in pure, hopeful white lies Germany.

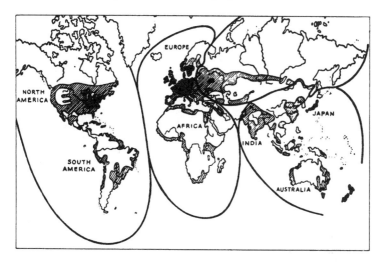

FIGURE 7.16. "Spheres of Influence" (*Facts in Review* 3, no. 13 [10 April 1941]: 182).

In trying to persuade the United States to remain neutral, Nazi cartographic propagandists flattered both isolationism and Monroe Doctrine militarism. Titled "Spheres of Influence," figure 7.16 uses bold lines to send a clear message to Americans: stay in your own hemisphere and out of Europe. Faintly resembling the lobes of Goode's interrupted projection (fig. 2.6), familiar to many students, the map also supported a geopolitical theater for Germany's Pacific ally, Japan. How successful the Nazi cartographic offensive might have been is moot, for the United States entered the war on the side of England after Japan attacked Pearl Harbor, Hawaii, on 7 December 1941.

Arrows, Circles, Place-Names, and Other Cartographic Assault Weapons

Few map symbols are as forceful and suggestive as the arrow. A bold, solid line might make the map viewer infer a well-defined, generally accepted border separating neighboring nations with homogeneous populations, but an arrow or a set of arrows can dramatize an attack across the border, exaggerate a concentration of troops, and perhaps even justify a "preemptive strike." As figure 7.17 demonstrates, arrow symbols

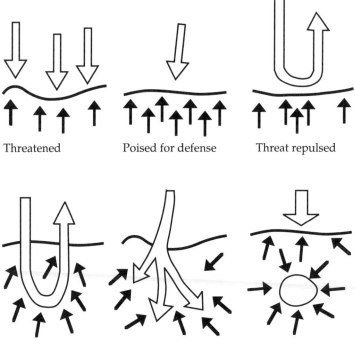

FIGURE 7.17. Arrow symbols portraying a variety of maneuvers and stalemates.

can vary in size, number, and arrangement to portray a range of military confrontations, from overwhelming threats and courageous standoffs to invasions with varying degrees of success. During World War II and the Korean War, many American newspapers used daily battlefield maps with forceful and suggestive arrows to give their readers a generalized blow-by-blow account of the Allied forces' victories and defeats. As figure 7.18 demonstrates, prominent arrows and black areas portraying captured territory could dramatize the threat of an advancing enemy.

A less abstract cousin of the arrow is the bomb or missile symbol. Everybody knows what it is and fears its referent. Lines of miniature missiles and stacks of ominous little red or black bombs readily impress map viewers with the comparative sizes of opposing arsenals. Orientation is also important:

bombs are stockpiled horizontally but dropped vertically, whereas missiles are stored upright but hurled horizontally. To justify an expanding defense budget, a propagandist might even stage a mininuclear attack, complete with a victorious response. Maps can even make nuclear war appear survivable.

The specter of nuclear warfare sends threatened nations and pacifists worldwide to the cartographic arsenal for an honored piece of geopolitical ordnance, the circle. Diplomats and military strategists have found the circle particularly useful in showing the striking zones of aircraft, and modern strategists find circles indispensable when discussing the range of guided missiles. Circles bring to the map a geometric

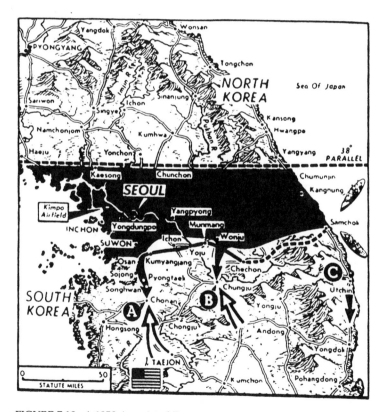

FIGURE 7.18. A 1950 Associated Press newspaper map uses black shading to mark the part of South Korea invaded by North Korean forces and arrows to portray troop movements.

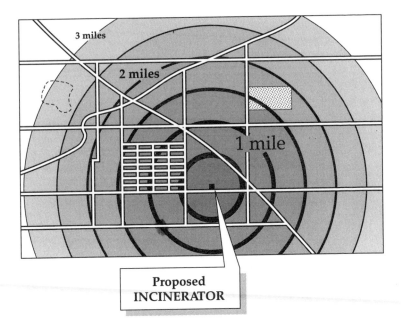

FIGURE 7.19. A local environmental protection group might seek to arouse citizen support with a propaganda map on which concentric circles have progressively more threatening labels closer to the site of a proposed incinerator.

purity easily mistaken for accuracy and authority. Yet on few small-scale maps do circles on the sphere remain circles in a two-dimensional plane. Even local environmental activists find circles useful, especially when arranged concentrically around the site of a proposed incinerator or nuclear power plant, and with ever larger, more threatening labels for closer circles, as in figure 7.19 (pl. 7).

Naming can be a powerful weapon of the cartographic propagandist. Place-names, or *toponyms*, not only make anonymous locations significant elements of the cultural landscape but also offer strong suggestions about a region's character and ethnic allegiance. Although many maps not intending a hint of propaganda might insult or befuddle local inhabitants by translating a toponym from one language to another (as

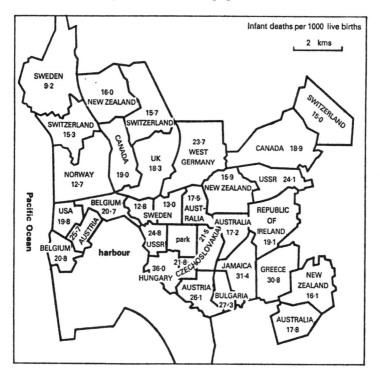

FIGURE 7.20. Dramatic map comparing infant mortality rates for parts of San Diego, California, with national rates of various countries.

from Trois Rivières to Three Rivers) or by attempting a pho-
netic transliteration from one language to another (as from
Moskva to Moscow) and even from one alphabet to another
(as in Peking or Beijing), skillful propagandists have often
altered map viewers' impressions of multiethnic cultural
landscapes by suppressing the toponymic influence of one
group and inflating that of another.

Local social activists can also use the suggestive power of
place-names to make a point cartographically. Figure 7.20,
for instance, is an infant mortality map of San Diego, California, that strongly indicts intraurban inequalities in maternal
and infant health care. As the map notes, some parts of the
city are comparable to highly developed western European
nations such as Sweden and Switzerland, whereas other
neighborhoods are similar to Hungary or Jamaica. Figures

7.19 and 7.20 both demonstrate that cartographic propaganda can be an effective intellectual weapon against an unresponsive, biased, or corrupt local bureaucracy. Like guns and lacrosse sticks, maps can be good or bad, depending on who's holding them, who they're aimed at, how they're used, and why.

Chapter 8

MAPS, DEFENSE, AND DISINFORMATION: FOOL THINE ENEMY

Compared with military maps, most propaganda maps are little more than cartoons. A good defense establishment knows how to guard its maps and their geographic details and yet at times to leak false information the enemy might think is true. Providing some accurate information is necessary, of course, if the "disinformation" is to be credible. An intellectual weapon in political propaganda, the map is a fundamental tactical weapon for military counterintelligence and covert diplomacy.

This chapter addresses how and why governments guard maps, hide geographic information, and sometimes even distribute deliberately falsified maps. The first section discusses the very real need for cartographic security, the second examines the now-admitted excesses of Soviet cartographers who deliberately doctored their maps, and the third section explores how governments sometimes mislead their own citizens by failing to include threats to a sound environment and other possible embarrassments.

Defense and a Secure Cartographic Database

No doubt about it: mapped information often must be guarded. If knowledge is power, an enemy's knowledge of your weaknesses and strengths is a threat. Maps can also betray your plans, as Giovanni Vigliotto discovered. In 1981 an Arizona jury found this fifty-three-year-old ladies' man guilty of fraud and bigamy. Giovanni, who claimed to have married more than 105 women over thirty-three years, invariably cut short the honeymoon by absconding with his victim's cash and jewelry. Had he not left behind an annotated map when he abandoned one of his wives, Giovanni might not have been caught.

Nations too try to keep maps out of enemy hands, even obsolete maps. In 1668, Louis XIV of France commissioned three-dimensional scale models of eastern border towns, so that his generals in Paris and Versailles could plan realistic maneuvers. On exhibit in Paris in La Musée des Plans-Reliefs, at the Hôtel des Invalides, these highly detailed wood and silk models are amazingly accurate portraits of seventeenth-century French towns. As late as World War II, the French government guarded them as military secrets with the highest security classification. Less cartographically ambitious states also tightly control battle plans and maps giving strategic information about communications, fortifications, and transport. Giving the enemy a detailed map has often been considered an act of treason, unless of course the map itself is a fraud designed to confuse the opponent or persuade him to attack or not attack.

What is mapped as well as the maps themselves must be kept confidential, for to reveal an interest in a particular area or features is to reveal one's plans. Governments compile maps of foreign areas in secure, windowless buildings using cartographers with "secret" or "top secret" security clearances because they don't want enemies, neutral countries, or even allies to know what interests them.

Mapping agencies must, of course, guard against fire, natural disasters, and sabotage, but computers, modern electronic telecommunications, and nuclear weapons call for special cartographic security. In the 1970s electronic maps stored in computer databases began to replace traditional maps on paper or plastic film as the principal format for organizing and storing geographic data. Highly efficient for ready updating and rapid retrieval, electronic maps are also vulnerable to computer hackers and to a type of nuclear attack called *electromagnetic pulse*. Unknown before telecommunication networks allowed computers to talk to each other as well as to distant users, the hacker is a compulsive computer enthusiast who likes to enter and alter others' databases. Malicious hackers sometimes alter information or develop so-called computer viruses that destroy data and programs. Known to have changed school grades and bank balances and to have penetrated supposedly secure Department of Defense computers, hackers must be intrigued by the challenge and thrill of entering a cartographic data system and moving Montana or flooding the Dead Sea.

A second threat to maps stored on magnetic media, the electromagnetic pulse (EMP), is a burst of radiation hurled downward by a high-altitude thermonuclear explosion. Able to destroy power and telecommunications transmission systems and damage integrated circuits, optical fibers, and magnetic storage devices, this sudden, intense radiation could render unreadable most maps recorded on tapes and diskettes. To protect themselves against EMP, governments are attempting to "harden" their electronic information systems and to stockpile maps on paper, microfilm, and other nonmagnetic media. Should civilization survive a nuclear attack, the traditional, bulky cartographic image will be the ultimate backup for the more flexible yet more vulnerable electronic map.

Soviet Cartography, the Cold War, and Displaced Places

Keeping all or even most maps away from enemy eyes is a nearly impossible task, especially in the age of the printing press, the electrostatic copier, and the computer. Maps are easily mass-produced or copied, and intense security can impose a severe burden on a nation's own citizens and even its military. A still greater hindrance to map users, though, is deliberate, widespread cartographic "disinformation," a Cold War tactic of the USSR.

Although other nations might have intentionally distorted their maps, the Soviet Union's systematic falsification of geographic location is now a well-known part of the recent history of cartography. In the late 1930s, after the NKVD, or security police, assumed control of mapmaking, the Soviet cartographic bureaucracy began to deliberately distort the position and form of villages, coastlines, rivers, highways, railroads, buildings, boundaries, and other features shown on maps and atlases sold for public use. This policy reflects a police-state mentality not unlike the misguided disinformation campaigns and cover-ups that at times have embarrassed the United States, Britain, and other Western governments. Ironically, the

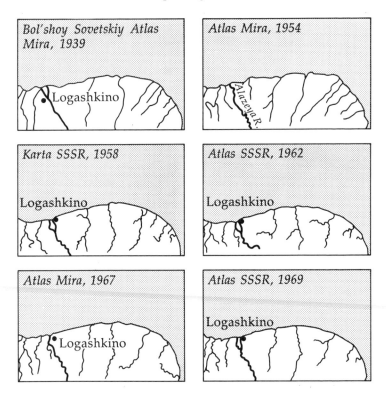

FIGURE 8.1. Representation of Logashkino and vicinity, on the East Siberian Sea, on various Soviet maps published between 1939 and 1969.

Soviets accelerated their map distortions in the mid-1960s, when the United States had begun to deploy sophisticated spy satellites. Apparently they sought to make their maps be considered untrustworthy among military planners in China and the West, so that their opponents would not make electronic maps useful for guiding cruise missiles by recording coordinates from Soviet maps.

How severe were the distortions? Figure 8.1 provides an example in the puzzling peregrinations of the town of Logashkino, on or near the shore of the East Siberian Sea, as portrayed on various Soviet maps published between 1939 and 1969. The *Bol'shoy Sovetskiy Atlas Mira*, published in 1939, showed Logashkino on the Alazeya River well inland from the coast, but the *Atlas Mira* published in 1954 omitted it altogether and

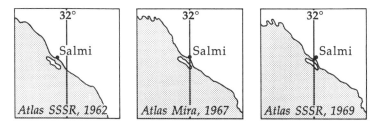

FIGURE 8.2. Representation of Salmi and vicinity, on Lake Ladoga near 32° E, on Soviet maps published between 1962 and 1969.

showed the river with only a single channel. *Karta SSSR* for 1958 recreated Logashkino on the coast and restored the river's other channel. *Atlas SSSR* published in 1962 offered a similar portrayal of the town and the river, but *Atlas Mira* for 1967 eliminated the eastern channel and moved Logashkino inland. Finally, the 1969 edition of *Atlas SSSR* again moved the town to the coast and showed both channels. The anonymous writer who described these mercurial representations for the magazine *Military Engineer* wryly observed, "Apparently there is such a town, but whether it is on the seacoast or on a river, or neither, is a matter of uncertainty when based on the work of the Soviet cartographers."

Displacement was particularly noticeable for towns near a meridian or parallel. Figure 8.2 illustrates the coordinate shifts for Salmi, a town on the north shore of Lake Ladoga, and a nearby offshore island. The heavy vertical line through each panel is the meridian 32° E. Although the *Atlas SSSR* showed Salmi about 10 km west of the meridian in 1962, the *Atlas Mira* edition of 1967 had moved the town eastward, directly astride the meridian. And *Atlas SSSR* for 1969 shifted the town 4 km farther east. The 32d meridian, which on the 1962 map fell well east of the island, nearly cut the island in half on the 1969 map. Although Salmi moved only 14 km, or nearly 9 miles, between 1962 and 1969, Soviet maps have displaced other towns by as much as 25 miles.

Soviet cartographic disinformation even affected tourist maps of urban areas. Detailed street maps of Moscow and other Soviet cities often failed to identify principal thoroughfares and usually omitted a scale, so that distances were difficult to estimate. Although local citizens were well aware of its presence,

Soviet street maps of Moscow suppressed the imposing KGB building on Dzerzhinski Square, as well as other important buildings. In contrast, the CIA pocket atlas used by American foreign service personnel has a detailed, fully indexed, easy-to-use map that shows the KGB headquarters and other important landmarks. Imagine the outcry if Soviet diplomats had better maps of Washington, D.C., than government workers, tourists, and other citizens.

So why did Soviet cartographers stop fudging their maps? First, cartographic disinformation is costly both to mapmaking and to economic development. To be convincing, feature falsification takes time and personnel better used to make the nation's maps more accurate and up to date. Maintaining a second, secure set of accurate maps is expensive, and providing carefully controlled access to economic planners and other decision makers is costly, time consuming, and risky. Second, spy satellites had made fudged maps less useful and surely less necessary than in the 1960s and earlier, when the USSR's enemies still depended largely upon existing old maps, spies, defectors, and U-2 spy planes. Modern intelligence satellites provide routine land-cover surveillance of potentially hostile countries, and image-analysis computer systems scan satellite imagery in a vigilant search for suspicious changes in terrain. Some intelligence satellites with high-resolution sensors can even alter their orbits to collect highly detailed imagery of suspect areas. Resolution adequate for capturing the numerals on a license plate is more than adequate for monitoring missile launch sites and troop movements. Indeed, a geopolitical rival's maps of parts of a country might be better than those available to its own citizens.

Features Not Shown, Maps Not Made

Are other countries' maps more open, more revealing, more complete than Soviet maps? Generally, yes. For example, figure 8.3, a large-scale state highway department map sold to the public, shows considerable ground detail for Griffiss Air Force Base in the city of Rome, New York. Home to long-range bombers of the Strategic Air Command (SAC) and advanced research projects on electronic aerial reconnaissance at the Rome Air Development Command (RADC), Griffiss would be a prime target in a war with the USSR. Yet cartographic openness is not universal in the United States, although the exceptions are often more puzzling than frustrating. As an illustration, figure 8.4 shows how maps of Catoctin Mountain National Park in western Maryland weakly camouflage Camp David, the famous presidential retreat, with the nondescript label "Camp 3." Yet surprisingly, the map also offers a few clues, such as the perimeter road along the security fence. As

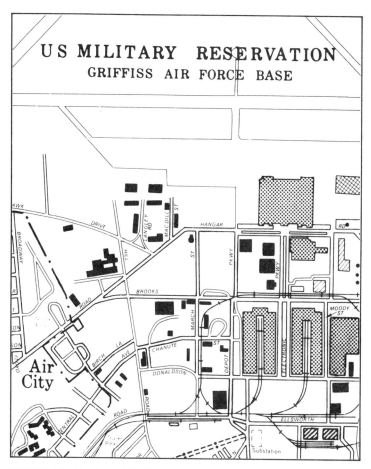

FIGURE 8.3. Portion of Griffiss Air Force Base, in Rome, New York, as shown on a large-scale planimetric map sold to the public by the state of New York.

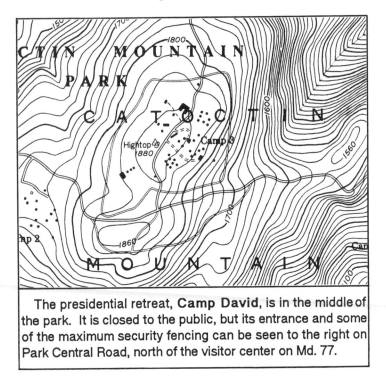

The presidential retreat, **Camp David**, is in the middle of the park. It is closed to the public, but its entrance and some of the maximum security fencing can be seen to the right on Park Central Road, north of the visitor center on Md. 77.

FIGURE 8.4. Camp David, a portion of Catoctin Mountain National Park, in Frederick County, Maryland, appears anonymously as "Camp 3" on a U.S. Geological Survey topographic map of the area (above). In contrast, a guidebook provides concise directions (below).

quoted at the bottom of figure 8.4, a guidebook for highway travelers in Maryland readily thwarts this feeble attempt at geographic anonymity by providing accurate directions to the camp.

Britain seems to be more cartographically paranoid than the United States, according to the *New Statesman*, a London journal of left-leaning political commentary. Britain's national civilian mapping agency, the Ordnance Survey, apparently maintains lists of sensitive sites that must be omitted or disguised on maps and somehow camouflaged on aerial photographs. Thus nuclear bunkers might masquerade as warehouses, whereas radio stations, plants processing nuclear fuel, and government-owned oil depots vanish altogether. Less devious perhaps, some countries, such as Greece, publish maps with large, telltale blank areas.

Yet Americans need not applaud themselves for their comparative openness. After all, British maps tend to show more detail at larger scales than American maps. Moreover, American maps often omit information that might embarrass industrial polluters or local officials. For example, figure 8.5 shows two maps centered on the infamous Love Canal, a neighborhood near Niagara Falls, New York, contaminated by hazardous waste. The 1946 map, which shows the canal as a long, straight vertical feature, fails to indicate use of the canal since 1942 as a dump for chemical waste. The 1980 map not only shows no trace of the filled-in canal but ignores the area's tragic history: dumping continued until 1953; developers filled in the canal and built homes there in the 1950s; the city built a public elementary school across the filled-in canal in 1954; chemical seepage spread up to the surface and laterally into the basements of nearby homes. After people and pets became ill in the early 1970s, soil analyses revealed abnormally

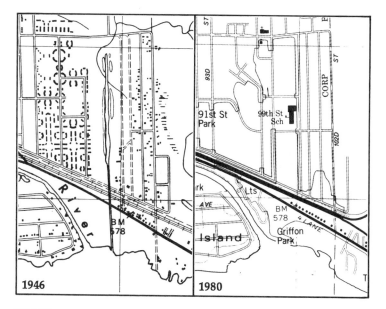

FIGURE 8.5. Love Canal area in Niagara Falls, New York, as shown on large-scale topographic maps from 1946 (left) and 1980 (right).

high concentrations of chlorobenzene, dichlorobenzene, and toluene; and in 1978 the New York health commissioner declared a state of emergency and relocated 239 families. Although both federal and state mapping agencies might contend that topographic maps should show only standardized sets of readily visible, more-or-less permanent features, such assertions seem hypocritical when these agencies' maps routinely include boundary lines, drive-in movie theaters, and other elements far less important to human health.

As historian of cartography Brian Harley has noted, government maps have for centuries been ideological statements rather than fully objective, "value-free" scientific representations of geographic reality. Harley observed that governments practice two forms of cartographic censorship—a censorship of secrecy to serve military defense and a censorship of silence to enforce or reinforce social and political values. Social-political censorship can assert the power of the state or the rights of private landowners, and it can also attempt to calm ethnic minorities, as in 1988 when New York governor Mario Cuomo decreed that all derogatory place-names be stricken from state maps. But as with the Love Canal maps, this second, more subtle form of cartographic censorship usually occurs as silences—as features or conditions ignored. Hence basic maps of most cities show streets, landmark structures, elevations, parks, churches, and large museums—but not dangerous intersections, impoverished neighborhoods, high-crime areas, and other zones of danger and misery that could be accommodated without sacrificing information about infrastructure and terrain. By omitting politically threatening or aesthetically unattractive aspects of geographic reality, and by focusing on the interests of civil engineers, geologists, public administrators, and land developers, our topographic "base maps" are hardly basic to the concerns of public health and safety officials, social workers, and citizens rightfully concerned about the well-being of themselves and others. In this sense, cartographic silences are indeed a form of geographic disinformation.

Chapter 9

LARGE-SCALE MAPPING, CULTURE, AND THE NATIONAL INTEREST

☀

Lest the reader be left with the impression that agencies responsible for large-scale maps are either deliberately devious or flagrantly irresponsible, this chapter explores the influence of culture on topographic mapping. Always a loaded concept for social scientists, culture accounts for many routine, seemingly automatic decisions about features and their portrayal as well as for national policy on collecting and disseminating cartographic information. As we'll see, our nation's topographic maps reflect widely shared Western values (geometric precision, consistency, completeness, cost-effectiveness) as well as the professional subculture of mapmaker-bureaucrats, whose traditions are rarely questioned. A few international comparisons demonstrate that the American way, however workable, is not the only way.

Focused on the operations and mores of national mapping organizations, in particular the U.S. Geological Survey, this chapter offers insights to occasional users of topographic maps as well as to businesses and local governments heavily dependent on cartographic information. After a cursory examination of the national mapping enterprise and its evolution, the first section looks at how rigorously defined feature types, map symbols, and generalization procedures promote consistency among the hundreds or thousands of individual quadrangle maps that constitute a map series. The second section, on cartographic agendas, explores mapmaking as an institutional process subject to bureaucratic norms, administrative red tape, and political pressure.

Standards and Specifications

Nothing better reflects the government cartographer's bureaucratic mentality than the standards and specifications for a nationwide series of topographic maps. In partitioning an entire country among a largely arbitrary grid of rectangular areas called *quadrangles,* the national mapping organization willingly sacrifices political, ethnic, and physical boundaries to the convenience of uniformly spaced meridians and parallels—a divide-and-conquer strategy that makes complete coverage seem both doable and essential. Even though the mapmaker is not a foreign invader, the interests of local communities divided among two or more quadrangles crumble in the path of a cartographic enterprise committed to scientifically higher, geographically broader needs.

What are these needs? National mapping as practiced in the United States and other developed countries has long served complementary goals of national defense and economic development. Military leaders need detailed maps to identify points of attack and plan their defense. Since an effective defense requires not only planning but rapid deployment of troops and weapons, a nation also needs roads, railways, and dependable waterways—infrastructure equally useful in getting raw materials to factories and manufactured goods to market. Topographic surveys, which help civil engineers lay out routes and build bridges, provide the geometric foundation for maps on which transportation structures are basic features. Maps, of course, also promote mineral exploration and agricultural development, as well as support private ownership—and taxation—of land. And as a consequence of economic development, industrialized societies need maps for environmental protection and growth management. However specific a nation's mapping requirements, most series of large-scale maps address several, if not all, of these four principal objectives.

A geographic windfall of sorts, large-scale maps afford a base for compiling a wide variety of smaller-scale products, including road maps, real-estate maps, and maps in textbooks, scientific papers, and newspapers. In 1892, Henry Gannett, the Geological Survey's first chief topographer, recognized the importance of detailed cartographic sources by calling his department's 1:125,000 and 1:62,500 quadrangle maps "mother maps." An admirer of the systematic large-scale map series produced by national surveys in Europe, Gannett sought a comparable treatment of the United States.

In selecting cartographic features, early USGS mapmakers

gave highest priority to "natural features" such as streams and landforms. Among the "artificial features" of the built environment, they distinguished "public culture," such as canals, railroads, and wagon roads, from "private properties," such as houses, fields, and orchards. Public topography was worth showing, but private topography, deemed less permanent, could require constant revision at great cost, as well as interfere graphically with symbols representing natural and public features. Even so, some private topography (dwellings, larger commercial buildings) seemed an essential part of the country's basic geographic frame of reference. And in the 1940s, a new series of 1:24,000 maps, based on 7.5-minute quadrangles each covering only a fourth as much territory as existing 1:62,000, 15-minute maps, allowed more room for showing homes, factories, sewage treatment plants, power lines, and other features collectively identified in cartographic jargon as "culture."

In addition to specifying which natural and built features to include, USGS officials prepared a manual to guide workers throughout the agency in both generalization and symbolic representation. Take, for example, the water gap, a deep, narrow break cut by a river through an otherwise continuous linear ridge. As illustrated in figure 9.1, water gaps provide efficient passage for closely spaced roads and railways. Crowded features challenge mapmakers to maintain clarity and legibility while packing a narrow graphic corridor with symbols portraying roads, railways, river banks, and their interior flow lines as well as contour lines describing slope and elevation. To promote consistency within a map series and prevent disagreement among workers and supervisors, national mapping organizations established rules for dealing with cartographically contentious situations. In the case of water gaps, a prescribed order of inking—culture first, then drainage, and then topography—not only avoided graphic congestion by displacing contours and streams but obviated awkward kinks in track and highway alignments.

Standards and specifications have become ever more exacting, with rules devised by committees telling mapmakers how to select features and show details. As the following USGS road-selection rules illustrate, size is significant, but relevance to humans can make small features important:

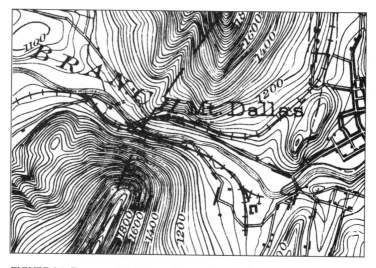

FIGURE 9.1. Because visibility usually requires road and railway symbols to be wider than the map scale calls for, drainage symbols and contour lines have been displaced laterally to minimize graphic congestion on this 1902 1:62,500 USGS topographic map (enlarged to 200% of original scale) showing a water gap just west of Everett, Pennsylvania.

> Private roads, access roads, and driveways less than 500 feet (152.4 m) in length will not be shown unless of landmark value in areas of sparse culture.

> All streets in populated places will be shown regardless of length.

And as these guidelines for portraying railways point out, a feature's overall form can be more relevant than its constituents and their exact positions:

> Within the [railroad] yard, main-line through tracks are shown correctly placed, but other tracks are symbolized, preserving as much as possible the distinctive pattern presented by the yard.

> Spur tracks, sidings, switches, and storage tracks are mapped accurately as to length, but may be adjusted positionwise [that is, displaced] if the map scale and adjacent detail require it.

Although a thick book of dry rules no doubt takes some of the fun out of being a mapmaker, standardization lets the map user who learned to decode federal maps in one area transfer these skills with confidence to other parts of the country.

Cartographic guidelines have deep roots in military mapping. Founded in 1879, the U.S. Geological Survey inherited mapping personnel from the four great post-Civil War surveys of the American West. Lead by Clarence King, F. V. Hayden, John Wesley Powell, and George M. Wheeler, these exploratory surveys addressed a variety of scientific, military, and economic goals and paved the way for the settlement of vast public lands in the little-known territories west of the 100th meridian. Surveyors and engineers typically had some military experience, and their mapping methods reflected techniques taught at West Point and used in the Union Army and the Corps of Engineers. As a result, topographic mapping in the new Geological Survey reflected military practice, a tradition fostered in later years by civil service rules favoring war veterans and by executive orders requiring at least minimal coordination of civilian and military agencies with common interests.

Perhaps the most subtle example of military influence on civilian topographic maps is the green tint representing woodland. Whether an area with trees is colored green reflects neither botanical nor ecological criteria but a tactical necessity. According to USGS topographic specifications, mappable woodland is "an area of normally dry land containing tree cover or brush that is potential tree cover," that is, "the growth must be at least 6 feet (2 m) tall and dense enough to afford cover for troops." Traditional measurement units also influence cartographic specifications: for example, a small, isolated woods is shown only if it exceeds an acre in area, and a similar one-acre minimum controls the portrayal of clearings within a forest. European maps, by contrast, are typically affected by standards based on kilometers and hectares.

Although U.S. military surveys of the nineteenth century learned (and borrowed) much from their European counterparts, large-scale U.S. maps reflect uniquely American traditions in content and symbolization. A noteworthy example is our portrayal of railways with a thin line punctuated by short, uniformly spaced cross-ticks (fig. 9.2, left)—a cartographic icon so readily recognized by American map users that otherwise complete map keys often omit it. Is this a natural symbol for railroads? Hardly: despite the vague suggestion of a steel rail running atop perpendicular wooden ties, the symbol's inherent graphic logic was surely not self-evident to the autocrats who established standards for topographic maps in Europe

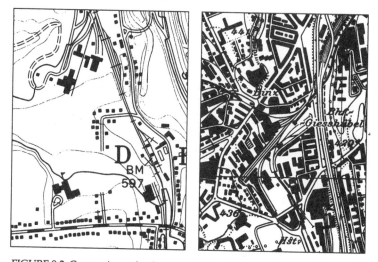

FIGURE 9.2. Comparison of railway symbols on large-scale topographic maps of the United States (left) and Switzerland (right). (Excerpts enlarged to 150% of original scale.)

and elsewhere. Consequently, American tourists on their first trip across the Atlantic sometimes experience a jarring but usually short-lived cartographic culture shock when they encounter the heavy, black, unadorned Eurostyle railroad symbols with few visual cues for viewers unaware, for instance, that on Swiss maps (as in the right panel of figure 9.2) *Bhf.* is an abbreviation for *Bahnhof* (meaning train station) and *Hst.* signifies a *Haltestelle* (literally a stopping place, or small station). As redundant or sole cues to meaning, abbreviations and feature names are valuable supplements to a purely geometric or pictorial graphic vocabulary.

Comparison of American and Swiss maps reveals numerous disparities in graphic conventions. As figure 9.3 illustrates, inherently ambiguous cartographic signs can have markedly different yet equally plausible meanings. To the Swiss, for instance, two thin parallel lines casing (enclosing) a thick dashed line represents a cog railway, with steep grades ascended by a locomotive or power car equipped with a rotary cog engaging a rack of evenly spaced vertical teeth positioned between the rails. But to American map readers, a similar symbol with a dashed red fill quite logically represents a secondary highway, graphically intermediate between a primary road

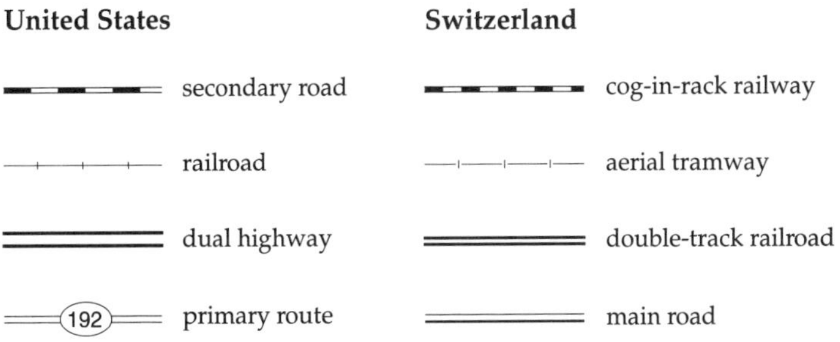

United States   **Switzerland**

━ ━━ ━━ ━ secondary road   ━ ━━ ━━ ━ cog-in-rack railway

─┼──┼──┼─ railroad   ─┼──┼──┼─ aerial tramway

═══════ dual highway   ═══════ double-track railroad

══(192)══ primary route   ═══════ main road

FIGURE 9.3. Comparison of selected line symbols used on American and Swiss topographic maps.

(shown by parallel black lines casing a solid red fill) and a light-duty road (similarly spaced black lines without the red fill). As another example, on Swiss maps a slight variation of the American railroad symbol portrays an areal tramway, in which a car suspended from a moving cable supported by two or more towers ascends a steep slope or crosses deep gorges. In contrast, American maps take the two thick solid lines showing a double-track Swiss railway, spread them apart, and color them red to represent a limited-access dual highway. Whereas American mapmakers rely on various highway shields (interstate, U.S. routes, state) to differentiate more important routes, the Swiss show thoroughfares with a double-line symbol noticeably thicker on the bottom or right. Despite these radical differences, the American and Swiss cartographic vocabularies work equally well in their own milieu because map readers either learn the code or use a key.

Visitors to Switzerland are frequently amazed by this small, prosperous, and compulsively neutral nation's exquisitely detailed topographic maps. Close inspection suggests that the 1:25,000 topographic maps of the Bundesamt für Landstopographie are much richer and more thorough than their 1:24,000 American counterparts. Maybe, maybe not. A large part of the obvious difference is the faint gray hill shading that supplements elevation contours on Swiss maps. No doubt about it: because of cartographic enhancement, the portrayal of Swiss terrain is not only easier to interpret than that of American terrain but a significant aesthetic enhancement. Another prominent difference is the detailed Swiss portrayal of buildings in urban areas (fig. 9.2, right). In contrast, American topographic maps blanket built-up areas with a light red tint and show only landmark buildings, such as courthouses and churches. Whereas the Swiss seem preoccupied with geometry and terrain, American topographers are like effusive local boosters obsessed with pointing out churches, schools, and other features not differentiated on Swiss maps. In addition to mildly idiosyncratic symbolic vocabularies, American and Swiss mapmakers have developed different notions of what's worth showing.

A comprehensive analysis of national mapping organizations would reveal surprising diversity in topographic vocabulary, with noteworthy differences between Canada and the United States, Austria and Germany, and Norway and Sweden. These differences have evolved somewhat like dialects, in response to necessity, isolation, and good ideas that catch on here but not there. Yet unlike verbal language, cartographic symbols are more rigorously controlled through formal standards and specifications administered by institutions that resist change and controversy. That's why our topographic maps as well as their Swiss counterparts overlook toxic waste dumps and evacuation zones around nuclear power plants.

Cartographic Agendas

Whereas standards and specifications spell out much of the government mapmaker's duties, national mapping organizations typically pursue subtle strategies evident only through a careful reading of their products and activities. An exploration of these hidden cartographic agendas reveals a bureaucratic mentality eager to appear productive and sometimes willing to cut corners.

Well before the multicultural concept of hate speech, mapmakers were aware that place-names like "Niger [sic] Creek" (fig. 9.4, left)—even if misspelled—were racially offensive. Naively, if not innocently, late-nineteenth and early-twentieth-century topographers striving for accuracy recorded these place-names as matter-of-fact representations of a geography authored by a politically incorrect generation of early Euro-American settlers, who might have intended no overt ill will in naming a feature Chink Creek, Dago Pass, Jap Gulch, or Nigger Lake. After World War II, as state legislatures and the Congress banned racial discrimination in schools, public places, housing, and hiring, racial epithets—at least in official, public discourse—began to reflect more harshly on the speaker than on the target. And so the nation's large-scale maps became a time bomb of outrage and embarrassment, with cartographic bureaucrats reluctant to act because of their own rigid rules.

As with everything else on topographic maps, naming and renaming physical features had become a formal process, overseen in this case by the Domestic Names Committee of the U.S. Board on Geographic Names, which spends much of its time reviewing recommendations from its state-level counterparts. Despite official channels available to anyone offended by a name on a map, getting rid of one name usually meant replacing it with another. Sounds simple, but documenting an acceptable replacement name can be a contentious, time-consuming process. Few offensive names disappeared through direct appeal to the board.

Eventually the federal government did act, but only at the request of the Department of the Interior, which includes the Geological Survey. Separate directives called for changing all occurrences of the two most worrisome slurs to comparatively neutral synonyms. In 1962—less than a decade before "black"

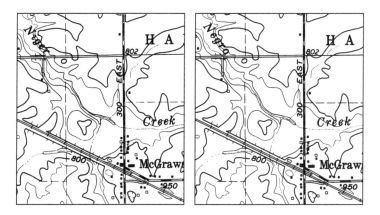

FIGURE 9.4. "Niger Creek" on a 1980 edition of the Bunker Hill, Indiana, 7.5-minute topographic map (left) was renamed "Negro Creek" on a 1994 revision (right). Had it been spelled with two *g*s, the racial slur might have been identified earlier when the Geological Survey searched its electronic geographic names database for offensive toponyms.

became the preferred term for "African-American"—Interior Secretary Stuart Udall ordered the substitution of "Negro" for "Nigger" whenever an offending map sheet was revised and reprinted. (Oddly, the example in figure 9.4, although corrected in a 1994 revision, had been overlooked in a 1980 edit.) And in 1967, a similar order replaced the pejorative "Jap" with the innocuous "Japanese." For whatever reason, Chinese-Americans and Hispanic-Americans have yet to win blanket eradication of "Chink" and "Dago."

While liberals wring their hands over the mapmaker's do-it-by-the-book intransigence over racial slurs, fiscal conservatives might object with equal vigor to the perfectly executed, largely blue maps of wholly inundated quadrangles in the middle of the Great Salt Lake. A case in point is the Rozel Point SW, Utah, 7.5-minute quadrangle, which derives its name from a land feature on the next row up, one sheet to the right. Except for the titles, grid lines, and marginal notations similar to those on maps of Fresno and Kalamazoo, the 1:24,000 Rozel Point SW sheet is a featureless light-blue rectangle adorned only by a note at the center

ELEVATION 4200

providing the approximate elevation of the water surface above sea level but revealing nothing about the water's depth. A round number is appropriate because the Great Salt Lake, like other water bodies in the Great Basin, rises and falls in response to rain, snowmelt, and evaporation. Nice to know, though, that a topographic map is ready for revision should the lake dry up and reveal some real topography.

Two other, more plausible explanations exist, neither particularly reassuring. Taking the high road, mapmakers can argue that featureless quadrangles are quadrangles nonetheless, and that completeness demands maps for all inland quadrangles, regardless of content. After all, who knows when an obsessive limnologist might want to decorate an atrium with a collage of topographic maps showing the entire Great Salt Lake. And at a more venal level, a production-conscious supervisor might cleverly enhance the annual report by adding at low cost to the tally of "quadrangles completed." Because the first motive provides a plausible cover for the second, a joint explanation is likely.

If this portrait of a cost-conscious federal bureaucracy

addicted to completeness seems farfetched, consider the "provisional maps" introduced in the early 1980s. Pressured by the White House to cut costs yet eager to complete the 7.5-minute series, Geological Survey managers reviewed their multistep topographic mapping production process. An evaluation of different strategies suggested that the average number of person-hours per quadrangle could be trimmed from 745 to 573 by increasing the amount of preliminary research and postsurvey compilation while cutting back on field survey, graphic artwork, and editing. Though they might not look as nice, the new maps would still meet national map accuracy standards. And the definition of "provisional" as "prepared beforehand or temporary, to be followed by something permanent" promised the few irate map connoisseurs a more aesthetically satisfactory product at an unspecified later date.

How aesthetically deficient are these economy-grade maps? Let the excerpt in figure 9.5 speak for itself. Although cartographic beauty is forever in the eye of the beholder, scratching on the negative with a sharp instrument to make faint, jerky labels like "Boat Ramp" and "Episcopal Ch. Spire" is a Neanderthal approach to what the Geological Survey calls "map finishing." Although labels identifying minor landmarks, locally important facilities, and survey markers (like the 4-foot tide-level benchmark coded "BM 4 – Tidal") are visually recessive, scratch lettering reflects an appalling disdain for style, grace, and legibility. And as a consequence of less editing and difficult corrections, provisional maps are more vulnerable than conventional topographic maps to spelling errors.

Ironically, about the time its drafters regressed to scribble, the Geological Survey tarted up the 7.5-minute provisional maps and a new series of 1:100,000 maps with a distinctive typeface named Souvenir. For an example of this distinctiveness, compare "Stoddartsville" in figure 9.6 with the more traditionally rendered "McGrawsville" in figure 9.4. Note, in particular, the stubby serifs (short strokes at the ends of letters) and distinguishing curved diagonal strokes in the lower-case letter v as well as in capital letters such as K, U, and Z. Advertising art directors liked this warm, contemporary look, and Pizza Hut, a national fast-food chain, adopted Souvenir for its ads and menus. Two decades after Madison Avenue embraced Souvenir, federal mapmakers cut themselves a slice.

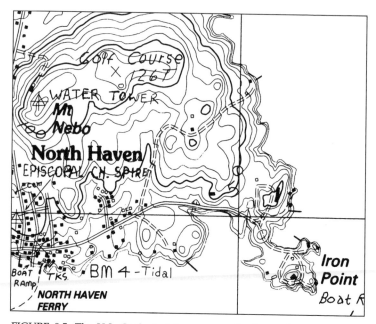

FIGURE 9.5. The U.S. Geological Survey expedited completion of its 7.5-minute, 1:24,000 series by scratching crude labels on photographic negatives of "provisional edition" maps, such as the North Haven East, Maine, sheet (enlarged to 150% of original scale).

What Geological Survey managers were up to no one can fathom with certainty. Perhaps they wanted to give their product a contemporary, user-friendly appearance, thereby promoting sales at the expense of the topographic map's traditional aura of authority and power. Maybe a top-level manager—a pizza lover, of course—saw Souvenir, liked it, and encouraged subordinates to use it. Possibly a design consultant with a background in advertising recommended it. In a more cynical interpretation, though, USGS officials might have recognized the power of bold, thick labels like "North Haven" in figure 9.5 to draw attention from the emaciated lettering on their provisional maps. Whether by design or happenstance, this decoy effect is an appropriate form of face-saving for fast-food cartography.

Sacrificing appearance to save money can be serendipitous, as *photorevised* 7.5-minute topographic maps clearly demonstrate. To cope with keeping 54,000 topographic quadrangle

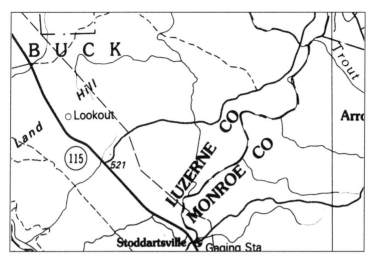

FIGURE 9.6. Many place-name labels on 1:100,000 series maps are set in Souvenir, as illustrated by the township (Buck), village (Stoddartsville), and county names on this excerpt (enlarged to 150% of original scale) from the 1986 planimetric (without contours) version of the Scranton, Pennsylvania, sheet.

maps reasonably up-to-date, the Geological Survey adopted a form of triage. Every five to ten years, on average, cartographers compare the most recent edition of a map sheet with new aerial photography in order to decide whether sufficient change has occurred to warrant revision and, if so, whether to draw an entirely new map or to produce a photorevised edition by merely adding new features such as roads and buildings in purple ink. The results look garish (especially if you don't like purple) as well as inconsistent (for instance, when contour lines around new limited-access dual highways are not revised to show cuts and fills), but the obvious advantage of more timely maps for many more quadrangles than otherwise possible outweighs these aesthetic shortcomings. And however inelegant, purple symbols usefully highlight areas and type of change as well as warn map users that static maps almost always lag behind the dynamic landscape they pretend to portray. All maps lie, but few are as frank as photorevised topographic sheets.

Committed to geometric accuracy and geographic completeness, federal topographers seldom make outrageous mistakes, but when they do, the images are memorable. The most

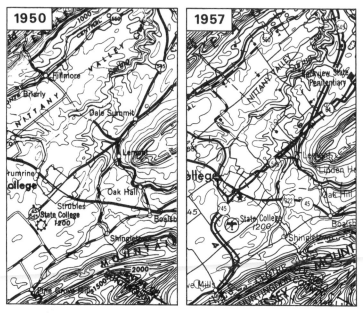

FIGURE 9.7. Excerpts from various editions of the 1:250,000 Harrisburg, Pennsylvania, map chronicle the puzzling attempt to abandon a nonexistent, erroneously inserted railroad running northeast from State College. (Maps reduced from 1:250,000 to approximately 1:330,000.)

outrageous example I know involves an equally outrageous cover-up. Figure 9.7 shows a nonexistent railway running northeast from State College, Pennsylvania, to Dale Summit and then turning north along state highway 545. This cartographic fiction appeared on the 1:250,000 Harrisburg sheet in a 1957 edition, compiled by Army Map Service personnel, who also worked on the 1:250,000 series. Although the compiler might have resurrected the short abandoned branch line from State College south to Pine Grove Mills from air photos or an old map, no railway ever ran northeast of town, through a business district and established residential neighborhood and across moderately steep terrain, which would have required at least one noteworthy trestle or an enormous fill—having lived there, I know. But that's only half the story. A 1965 edition with "limited revision by the U.S. Geological Survey" caught the error, but instead of removing the faulty railway symbol altogether, the cartographic artist substituted an abandoned rail-

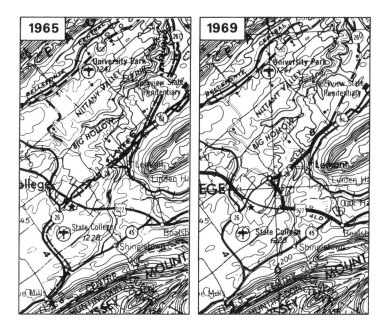

road, represented by an interrupted, dashed version of the tra-ditional cross-ticked railway symbol. By 1969, the federal topography factory finally got it right and dropped the ficti-tious feature once and for all.

Although gratuitous symbols usually reflect blunders, dele-tions typically follow careful deliberation of official feature standards. In the 1980s, for instance, when thoughtless sou-venir hunters and other looters became especially troublesome at historic sites in the southwest, the National Park Service asked the Geological Survey, the Automobile Club of Southern California, and other map publishers to remove ruins and other archeological sites from their maps and guidebooks. In complying, USGS mapmakers eliminated the petroglyphs (rock carvings) identified on the 1958 15-minute topographic sheet covering Ismay, Colorado (fig. 9.8, above), from the more detailed 7.5-minute map issued in 1985 (fig. 9.8, below). Most cultural sites attractive to treasure-hunters will no doubt disap-pear from maps as revision progresses, however slowly, but with blatant clues readily available in map archives, the policy of suppressing archeological features seems a bit late. Although the cartographic contribution to cultural looting is

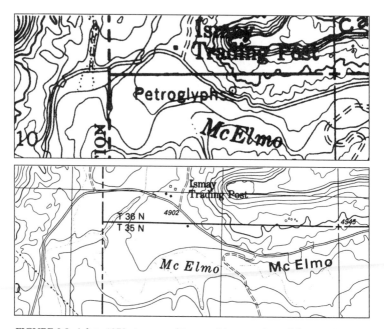

FIGURE 9.8. A late 1950s topographic map (above, enlarged from 1:62,500 to 1:24,000) pointed out the petroglyphs just southeast of the Ismay Trading Post, but more recent 7.5-minute maps (below) omit this and numerous other archaeological sites.

impossible to assess, accuracy and completeness clearly can have unintended consequences.

National mapping organizations are easy to criticize for much the same reason maps must lie: the enormous number of choices in selecting features, assigning symbols, and setting the scale of a map series. There's the added responsibility, though, of keeping tens of thousands of large-scale maps current through periodic review and revision. Because of limited resources, even a range of map series cannot satisfy the diverse requirements of map users. And while agencies cope with political pressures for greater cost-recovery, more effective federal-state-local coordination, and increased privatization, technological advances create ever more promising yet costly choices as well as a growing need to reassess priorities and revise standards. To be adequately informed, the map user must be at least vaguely aware of how cartographic bureaucracies work, what they value, and how values and biases affect their products.

Chapter 10

DATA MAPS:
MAKING NONSENSE OF THE CENSUS

A single set of numerical data, say, for the states of the United States, can yield markedly dissimilar maps. By manipulating breaks between categories of a choropleth map, for instance, a mapmaker can often create two distinctly different spatial patterns. A single map is thus just one of many maps that might be prepared from the same information, and the map author who fails to look carefully at the data and explore cartographic alternatives easily overlooks interesting spatial trends or regional groupings. Moreover, because of powerful personal computers and "user-friendly" mapping software, map authorship is perhaps too easy, and unintentional cartographic self-deception is inevitable. How many software users know that using area-shading symbols with magnitude data produces misleading maps? How many of these instant mapmakers are aware that size differences among areal units such as counties and census tracts can radically distort map comparisons? In addition to the ill-conceived charts of hacker-cartographers, wary map users must watch out for statistical maps carefully contrived to prove the points of self-promoting scientists, manipulating politicians, misleading advertisers, and other propagandists.

This chapter uses several simple hypothetical examples to examine the effects of areal aggregation and data classification on mapped patterns. Read it carefully and look closely at the maps and diagrams, and this excursion into cartographic data analysis should be richly rewarding rather than technically tedious. Anyone interested in public-policy analysis, marketing, social science, or disease control needs to know how maps based on census data can yield useful information as well as flagrant distortions.

Aggregation, Homogeneity, and Areal Units

Most quantitative maps display data collected by areas such as counties, states, and countries. When displayed on a map, presented on a statistical plot, or analyzed using correlation coefficients or other measures, geographic data produce results that reflect the type of areal unit. Because different areal aggregations of the data might yield substantially different patterns or relationships, the analyst should qualify any description or interpretation by stating the type of geographic unit used. Noting that values generally increase from north to south "at the county-unit level" warns the reader (and the mapmaker as well!) that a different trend might arise with state-level data, for instance.

Areal aggregation can have a striking effect on the mapped patterns of rates and ratios. A ratio such as the average number of television sets per household might, for example, produce radically different maps when the data are aggregated separately by counties and by the towns that make up these counties. The three town-level maps in figure 10.1 are spatially

Number of Televisions

1,000	100	50	100	50	100	50
200	100	200	100	200	100	200
100	200	100	4,000	100	200	100
200	400	200	400	200	400	3,000

Number of Households

2,000	200	100	200	100	200	100
200	100	200	100	200	100	200
100	200	100	4,000	100	200	100
100	200	100	200	100	200	1,500

Televisions per Household

0.5	0.5	0.5	0.5	0.5	0.5	0.5
1.0	1.0	1.0	1.0	1.0	1.0	1.0
1.0	1.0	1.0	1.0	1.0	1.0	1.0
2.0	2.0	2.0	2.0	2.0	2.0	2.0

FIGURE 10.1. Town-unit number tables showing number of televisions (top left), number of households (top right), and average number of televisions per household (bottom) for twenty-eight hypothetical towns.

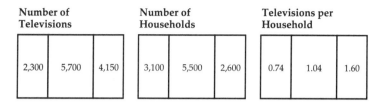

Number of Televisions			Number of Households			Televisions per Household		
2,300	5,700	4,150	3,100	5,500	2,600	0.74	1.04	1.60

FIGURE 10.2. County-unit number tables of number of televisions (left), number of households (middle), and average number of televisions per household (right) for a three-county aggregation of the twenty-eight hypothetical towns in figure 10.1.

ordered number tables, without graphic symbols, so that we can see how rate calculations depend on what boundaries are used and how they are drawn. The upper left-hand map shows the number of televisions in each of twenty-eight towns, the upper right-hand map represents the number of households, and the lower map portrays the television-ownership rate. Note the straightforward top-to-bottom pattern of the rates: low in the upper tier of towns, average in the two middle tiers, and high in the lower tier. Note also that three towns in the upper left, lower right, and just below the center of the region have relatively high numbers of households. These variations in household density underlie the markedly different left-to-right trend in television-ownership rates in figure 10.2, based on the same data aggregated by county.

Spatial pattern at the town-unit level of aggregation depends on how somewhat arbitrary political boundaries group towns into counties. Figure 10.3 uses two additional aggregations of these twenty-eight towns to demonstrate the possible effect of historical accident. The upper row of maps shows an alternative aggregation of towns into three horizontal counties that reflect the town-level top-to-bottom trend. In contrast, the lower series of maps shows an equally plausible aggregation into four counties, three based on the concentrations of households and one comprising the balance of the region. The television-ownership map for this lower set isolates what might be more urban counties from a single much larger, more rural county with an average of slightly more than one television per household. Graytone area symbols would yield very different choropleth maps for the three sets of rates shown in the right-hand maps of figures 10.2 and 10.3.

Another example illustrates how areal aggregation can affect geographic pattern. Whereas figure 10.3 demonstrates that different aggregations of towns into counties can yield markedly different county-level patterns, figure 10.4 illustrates how a single aggregation can produce the same county-level pattern from markedly different town-level patterns. Note that the town-level maps in figure 10.4 reflect a pattern of television-ownership rates very different from that in figure 10.1. Note in particular the progression of rates from a tier of low-ownership towns across the bottom of the region to a peak of much higher rates at the upper right. Yet when aggregated according to the county boundaries in figure 10.2, these data will yield similar county-unit rates. Comparing this trio of spatial number tables with those in figure 10.1 demonstrates the importance of stating clearly the data units used and of not assuming that a trend apparent at one level of aggregation exists at other levels as well.

The counties in these examples obviously are not homogeneous. But can we assume homogeneity even within the towns? What spatial variations in the distribution and density of these 11,200 households lie hidden in the network of town boundaries? Figure 10.5 presents one of many plausible point patterns that could produce the aggregated town-level counts and rates in figure 10.1. Three types of point symbols represent

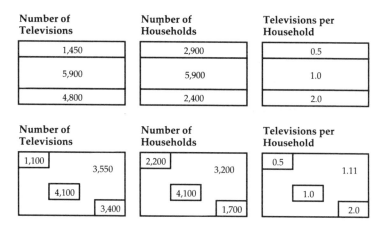

FIGURE 10.3. County-unit number tables based on other aggregations of the twenty-eight towns into counties.

Number of Televisions

190	285	200	350	350	210	890
455	450	1,085	960	895	520	1,260
355	315	525	480	595	360	700
130	120	80	100	80	110	100

Number of Households

100	150	100	200	100	50	100
350	300	700	600	500	200	300
500	450	700	600	700	400	500
650	600	400	500	400	550	500

Televisions per Household

1.90	1.90	2.00	1.75	3.50	4.20	8.90
1.30	1.50	1.55	1.60	1.79	2.60	4.20
0.71	0.70	0.75	0.80	0.85	0.90	1.40
0.20	0.20	0.20	0.20	0.20	0.20	0.20

FIGURE 10.4. Patterns of the number of televisions, number of households, and the television-ownership rate radically different from those in figure 10.1 could yield county-unit patterns identical to those in figure 10.2.

groups of 10, 100, and 500 households. Each symbol represents a group of households owning an average of 0, 1, or 2 televisions. The small, ten-household symbols represent rural residences, which lack TV receivers for religious reasons, lack of cable service, or a deep commitment to reading. Because of rough terrain, swamps, park or forest land, and undeveloped federal land, large parts of the region are uninhabited. Of the six large villages, with 400 or more households, two have two-TV households on the average, two have one-TV households, and two have video-free households. Although figure 10.5 contains elements of both the top-to-bottom town-level trend in figure 10.1 and the left-to-right county-unit trend in figure 10.2, its pattern of television ownership is more similar to the lower right of figure 10.3, where county boundaries segregate three large population clusters from the balance of the region. Yet even here the differences are striking, again demonstrating how the configuration of areal units can hide interesting spatial detail and present a biased view of a variable's geography.

Aggregation's effects become even more serious if the careless analyst or naive reader leaps from a pattern based on areal

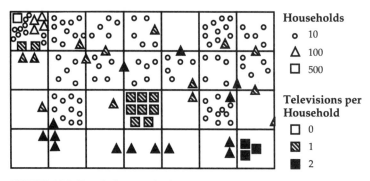

FIGURE 10.5. Detailed map of television ownership for villages and rural households illustrates one possible spatial structure that could yield the town-unit and county-unit maps in figures 10.1 and 10.2.

units to conclusions based on individual households. Consider, for instance, the large village toward the lower right-hand corner of figure 10.5. The average television-ownership rate here of 2.0 need not mean that each of the village's 1,700 houses has two TVs. Some households might have none while others have three or four or five. One or two residents might even be compulsive collectors, so that more than half the homes have one or none.

If households collecting old television sets seems farfetched, consider average household income, an index used frequently by social scientists and marketing analysts. Because of one or two innovative, unscrupulously manipulative, or otherwise successful residents, a small village might have an enormous *mean* household income. More a statistical quirk than a realistic reflection of overall local prosperity, this high average income might mask the employment of most villagers as household servants, gardeners, or security guards. Because nondisclosure rules prohibit a more precise publication of individual incomes, aggregated census data are the most refined information available. They provide an average for the place but say little about individual residents.

Are areally aggregated data bad? Surely not. In many cases, particularly in public policy analysis, towns and counties are the truly relevant units for which state and federal governments allocate funds and measure performance. And even more highly aggregated data can be useful, for instance, when governors and senators want to compare their states with the other forty-nine. Local officials and social scientists concerned with differences among neighborhoods readily acknowledge the value of geographic aggregation. Moreover, nondisclosure regulations needed to ensure cooperation with censuses and surveys require aggregation, and areally aggregated data are better than no data at all. Thus persons who depend upon local-area data encourage the Bureau of the Census to modify boundaries to preserve the homogeneity of *census tracts* and other reporting areas. And when tract data are not adequate, they sometimes pay for new aggregations of the data to more meaningful areal units.

What else can the conscientious analyst do? Very little aside from the obvious: know the area and the data, experiment with data for a variety of levels of aggregation, and carefully qualify all conclusions.

And what should the skeptical map user do? Look for and compare maps with different levels of detail, and be wary of cartographic manipulators who choose the level of aggregation that best proves their point.

Aggregation, Classification, and Outliers

Choropleth mapping further aggregates the data by grouping all areas with a range of data values into a single category represented by a single symbol. This type of aggregation addresses the difficulty of displaying more than six or seven visually distinct graytones in a consistent light-to-dark sequence. Often the mapmaker prefers only four or five categories, especially when the area symbols available do not afford an unambiguous graded series. (For aesthetic reasons or to avoid confusion with interior lakes or areas without data, black and white are not good graytone symbols for choropleth maps.)

But classification introduces the risk of a mapped pattern that distorts spatial trends. Arbitrary selection of breaks between categories might mask a clear coherent trend with a needlessly fragmented map or oversimplify a meaningfully intricate pattern with an excessively smoothed view. Figure 10.6 illustrates the influence of class breaks on the appearance of choropleth maps of the town-level television-ownership rates in figure 10.4. Note that the map on the left presents a

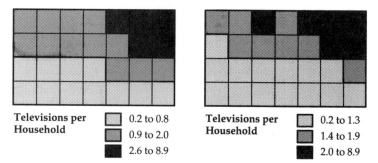

Televisions per Household
□ 0.2 to 0.8
▨ 0.9 to 2.0
■ 2.6 to 8.9

Televisions per Household
□ 0.2 to 1.3
▨ 1.4 to 1.9
■ 2.0 to 8.9

FIGURE 10.6. Different sets of categories yield different three-category choropleth maps for the data in figure 10.4.

clear, straightforward, readily remembered upward trend toward a peak at the upper right of the region, whereas the map at the right offers a more fractured view of the same data.

Classification raises many questions. Which map, if either, is right? Or if "right" sounds too dogmatic, which provides a better representation of the data? Don't both maps hide much variation in the broad third category, represented by the darkest symbol? Shouldn't the seven towns with rates of 0.2 occupy a category by themselves? Is a difference of, say, 0.1 at the lower end of the overall range of data values more important than a similar difference at the upper end? Can a three-class map provide even a remotely adequate solution?

But how many map authors bother to ask these questions? Because choropleth mapping is readily available through personal computers, so that the map viewer is often also the mapmaker, some instruction in map authorship is warranted.

Software vendors usually provide a few options for "automatic" classification, and naive mapmakers often settle for one of the easier options. Sometimes the computer program even provides a map instantly, without offering a choice of classification strategies. Called a "default option," this automatic choice of class breaks is a good marketing ploy because it gives the hesitant prospective purchaser an immediate success.

But does the default give you a good map? Figure 10.7 shows four-category mapped patterns produced by two common default classing options for the same town-level television-ownership data used in figure 10.6. The *equal-intervals* scheme, on the left, divides the range (8.7) between the lowest and highest data values (from 0.2 to 8.9) into four equal parts (each spanning 2.175 units). Note, though, that this classification assigns most of the region to a single category and that the third category (from 4.6 to 6.7) is empty. Of possible use when data values are uniformly distributed across the range, the only consistent asset of equal-interval classification is ease of calculation.

In contrast, the *quartile* scheme, on the right, ranks the data values and then divides them so that all categories have the same number of areal units. Of course, only an approximately equal balance is possible when the number of areas is not a multiple of four or when a tie thwarts an equal allocation (as occurs here at the upper left, where the highest category receives both of the towns with rates of 1.9). Although the map pattern is more visually balanced, the upper category is broad and highly heterogeneous, and the break between the second and third categories falls between two very close values (1.3 and 1.4). Yet the map based on these four quartile categories does have meaning for the viewer interested in the locations of towns in the highest and lowest quarters of the data values. Called *quintiles* for five categories and *quantiles* more generally, this rank-and-balance approach can accommodate any number of classes.

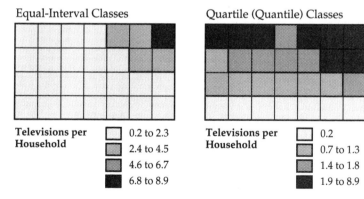

FIGURE 10.7. Two common classing schemes used as "defaults" by choropleth mapping software yield radically different four-category patterns for the data in figure 10.4.

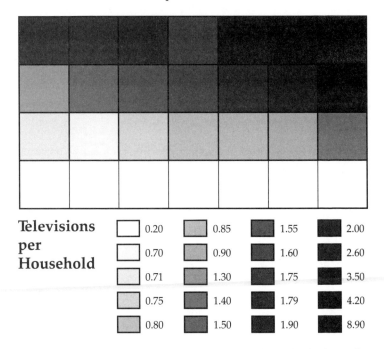

**Televisions
per
Household**

☐ 0.20	☐ 0.85	■ 1.55	■ 2.00
☐ 0.70	☐ 0.90	■ 1.60	■ 2.60
☐ 0.71	☐ 1.30	■ 1.75	■ 3.50
☐ 0.75	■ 1.40	■ 1.79	■ 4.20
☐ 0.80	■ 1.50	■ 1.90	■ 8.90

FIGURE 10.8. Continuous-tone, nonclassed choropleth map for the data in figure 10.4.

At least one mapping program offers the option of a "no-class" or "classless" choropleth map, on which each unique data value (perhaps up to fifty of them) receives a unique gray-tone. In principle this might seem a good way to sidestep the need to set class breaks. But as figure 10.8 illustrates, the gray-tones might not form a well-ordered series, and the map key is either abbreviated or cumbersome. Moreover, assigning each unique value its own category can destroy a clear, easily remembered picture of a strong, meaningful spatial trend. This ideal solution might not be so ideal after all.

Eschewing defaults and panaceas, the astute map author begins by asking two basic questions: How are the data distributed throughout their range? And what, if any, class breaks might have particular meaning to the map viewer? The answer to this second question depends on the data and on whether the map author deems useful a comparison with the national or regional average. On state-level maps, for instance, a break at the United States average would allow governors and senators to compare their constituents' or their own performance with that of the rest of the nation. Of course the map key would have to identify this break to make it truly meaningful.

After addressing the question of meaningful breaks, the conscientious map author might then plot a *number line* similar to that in figure 10.9. A horizontal scale with tick marks and labels represents the range of the data. Each dot represents a data value, and identical values plot at the same position along the scale, one above the other. The resulting graph readily reveals natural breaks, if any occur, and distinct clusters of homogeneous data values, which the classification ought not subdivide. Number lines allow the map author to visualize the distribution of data values and to choose an appropriate number of categories and appropriate positions for class breaks. Computer algorithms can also search the data distribution for an optimum set of breaks, but in many cases the computer-determined optimum is not significantly better than a visually identified suboptimal grouping. Rounded breaks and a more balanced allocation of places among categories can be important secondary factors in choropleth mapping.

Extremely high or extremely low values isolated from the rest of the distribution can confound both human cartographers and sophisticated mapping software. Should these *outliers* be grouped with markedly more homogeneous clusters higher or lower on the number line? Should each be accorded its own category? Can two or three widely separated data values at either end of the distribution be grouped into a single highly heterogeneous category? Or should each outlier be treated as its own category, with its own symbol, at the risk of

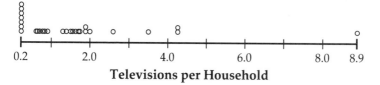

FIGURE 10.9. Number line for the town-level television-ownership rates in figure 10.4.

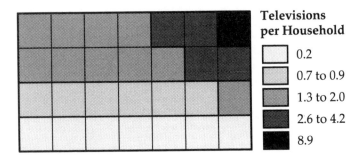

Televisions per Household

☐	0.2
☐	0.7 to 0.9
▨	1.3 to 2.0
▨	2.6 to 4.2
■	8.9

FIGURE 10.10. Choropleth map based on the number line in figure 10.9 and the character of the data.

reducing graphic differentiation among graytones? Or might the map author treat outliers as outcasts—errors or deviants that "don't belong"—and either omit them or give them a special symbol?

No simple, standard solution addresses all outliers. The map author should know the data, know whether these deviant values are real or improbable, and know whether a large difference between outliers really matters. Also important is the relation of outliers to the theme of the map and the interests of map viewers. For the television-ownership data in figure 10.9, an average of 8.9 TVs per household surely is not only exceptional but probably significantly higher than its neighboring values at 4.2. If not an error, it deserves special treatment in a category of its own. The next four lower values, 4.2 (twice), 3.5, and 2.6, might then constitute a single category; all are above the more plausible rate of 2.0, and yet 4.2 TVs per household is not improbable, especially in an affluent area.

Other breaks seem warranted between 0.9 and 1.3, a gap that includes the inherently meaningful rate of one TV per household, and between 0.2 and 0.7, to separate the seven videophobic towns at the lower end of the distribution. The resulting five-category map in figure 10.10 provides not only an honest, meaningful representation of the data values and their statistical distribution, but a straightforward portrayal of the spatial trend as well. An arbitrary classification, such as a computer program's default categories, is unlikely to do as well, even with six or more categories.

Classification, Correlation, and Visual Perception

Choropleth maps readily distort geographic relationships between two distributions. Hastily selected or deliberately manipulated categories can diminish the visual similarity of two essentially identical trends or force an apparent similarity between two very different patterns.

Consider as a case in point figure 10.11, a spatial data table and number line for the mean number of children per household, which has a strong town-level relationship to television ownership. Although the range of data values is not as broad for this index of family size, the highest values are at the upper right and the lowest values occur across the bottom of the region. Towns toward the right and toward the top of the region generally have more children in the home than do towns toward the bottom or left edge of the map. That the pair of maps in figure 10.12 shows identical spatial patterns for children and televisions is thus not surprising.

Statistical analysts commonly depict correlation with a two-dimensional scatterplot, with data values for one variable measured along the vertical axis and those for the other scaled along the horizontal axis. A dot represents each place, and the density and orientation of the point cloud indicates the strength and direction of the correlation. Figure 10.13 is a pair of scatterplots, both showing the strong positive association between the household rates for children and TVs. The perpendicular lines extending from the scales of the left-hand

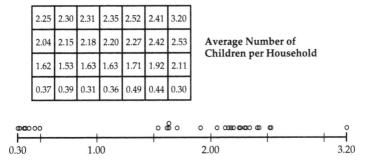

FIGURE 10.11. Spatial data table and number line for average number of children per household.

scatterplot into the scatter of points represent the class breaks in figure 10.12. These two sets of four lines each divide the scatterplot into an irregular five-by-five grid. Because all dots on the left-hand scatterplot lie within one of the five diagonal cells, the two five-category maps in figure 10.12 have identical patterns, enhancing the impression of a strong correlation.

Figure 10.13's right-hand scatterplot adds some carto-graphic skulduggery. As before, the perpendicular lines from

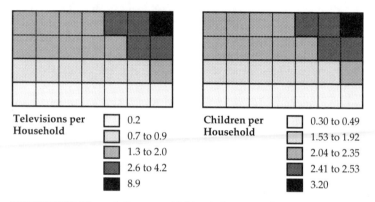

Televisions per Household

- ☐ 0.2
- ▨ 0.7 to 0.9
- ▨ 1.3 to 2.0
- ▨ 2.6 to 4.2
- ■ 8.9

Children per Household

- ☐ 0.30 to 0.49
- ▨ 1.53 to 1.92
- ▨ 2.04 to 2.35
- ▨ 2.41 to 2.53
- ■ 3.20

FIGURE 10.12. Choropleth maps with identical patterns for television owner-ship rate and average number of children per household.

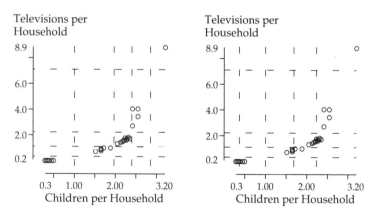

FIGURE 10.13. Scatterplots for the town-level television-ownership rate and average number of children per household. Additional lines on the left-hand scatterplot represent class breaks for the pair of maps in figure 10.12. Additional lines on the right-hand scatterplot show breaks used in figure 10.14.

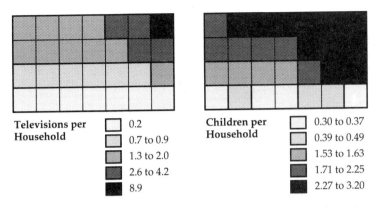

Televisions per Household
☐ 0.2
☐ 0.7 to 0.9
▨ 1.3 to 2.0
▩ 2.6 to 4.2
■ 8.9

Children per Household
☐ 0.30 to 0.37
☐ 0.39 to 0.49
▨ 1.53 to 1.63
▩ 1.71 to 2.25
■ 2.27 to 3.20

FIGURE 10.14. Distinctly different choropleth maps suggest minimal correlation between television ownership and family size.

the scales into the point cloud represent class breaks and form a five-by-five grid. But note that this configuration of breaks places all but four dots in an off-diagonal cell so that few towns will belong to the same category on both maps. Figure 10.14 demonstrates the resulting dissimilarity in map pattern and suggests a mediocre correlation at best. Similar tactics might make a weak relationship appear strong, especially if the maps are identical for the highest category, with the darkest symbol. Indeed, the spatial correspondence of the darkest, most eye-catching symbols strongly influences judgments of map similarity by naive map viewers. Some will even regard as similar two maps with roughly equal amounts of the darkest symbol—even if the high areas are in different parts of the region! Different area symbols for the two maps and different numbers of categories are other ways of tricking the map viewer or deluding oneself.

Another visual distortion might lie in the base map the data are plotted on. Not all sets of areal units are as uniform and visually equivalent as the square towns in the preceding examples. Figure 10.15 demonstrates this point with a deceptively similar-looking pair of maps based on the numerical data and class breaks of the visually dissimilar maps in figure 10.14. These twenty-eight towns vary markedly in size, and similarity is high because the largest towns belong to the same category. Towns not in the same category on both maps are smaller and less visually influential.

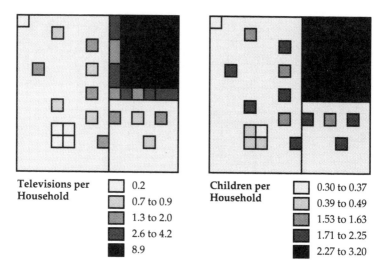

Televisions per Household		
	☐	0.2
	☐	0.7 to 0.9
	▨	1.3 to 2.0
	▨	2.6 to 4.2
	■	8.9

Children per Household		
	☐	0.30 to 0.37
	☐	0.39 to 0.49
	▨	1.53 to 1.63
	▨	1.71 to 2.25
	■	2.27 to 3.20

FIGURE 10.15. Similarity among large areas can distort visual estimates of correlation by masking significant dissimilarity among small areas. Numerical data and mapping categories are identical to those for the more obviously dissimilar pair of maps in figure 10.14.

Although this example is contrived, it is not atypical. Wards, census tracts, congressional districts, and other areal units designed to have similar populations often vary widely in area because of variations in population density. Disparities are even worse on county-unit maps, where populous metropolitan counties often are much smaller than rural counties with few inhabitants. The careful map viewer never judges numerical correlation by the similarity in map pattern alone and is especially cautious when some data areas are much bigger than others.

To avoid estimates of correlation biased by the size of areal units, the astute analyst will inspect the more egalitarian scatterplot, on which identical dots represent each area. As figure 10.16 illustrates, the density and orientation of the point cloud reflect the strength and direction of the correlation. If a straight line provides a good generalization of the point cloud, the correlation is called linear and the scatter of points around the line indicates the strength of the *linear correlation*. Positive relationships slope upward to the right, negative relationships slope downward to the right, and a point cloud without a discernible relationship has no apparent slope. Weak correlations have a wide, barely coherent scatter about the trend line, whereas for strong linear correlations most points are near or on the line. Not all correlations are linear, though; a strong *curvilinear correlation* has a marked curved trend, which a curved line fits better than a straight line.

Statisticians use a single number, the *correlation coefficient*, to measure the strength and direction of a linear correlation. Represented by the symbol r, the correlation coefficient shows the direction of the relationship by its sign and the strength of the relationship by its absolute value. The coefficient ranges from +1.00 to –1.00; r would be .9 or higher for a strong positive correlation, –.9 or lower for a strong negative correlation, and close to zero for an indeterminate or very weak correlation. (As a rule of thumb, squaring r yields the proportion of one variable's variation accounted for by the other variable. Thus, if r is –.6, the correlation is negative and one variable might be said to "explain" 36 percent of the other variable. A correlation coefficient measures only association, not causation, which depends upon logic and supporting evidence.)

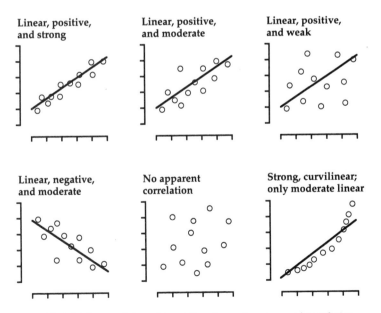

FIGURE 10.16. Scatterplots and trend lines for various types of correlation.

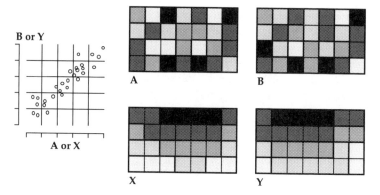

FIGURE 10.17. Two pairs of variables with identical scatterplots, correlation coefficients (r = .93), and class breaks, yet distinctly different map patterns.

Maps, scatterplots, and correlation coefficients are complementary, and the analyst interested in correlation relies on all three. The correlation coefficient, which provides a concise comparison for a pair of variables, measures only linear correlation. Yet a scatterplot quickly reveals a strong curvilinear relationship, with a mediocre value of r. Scatterplots also show outliers, which can greatly bias the calculation of r. But reliance upon visual estimation makes scatterplots poor for comparing strengths of relationships. Moreover, scatterplots and correlation coefficients tell us nothing about the locations of places, whereas maps, which present spatial trends, can offer unreliable estimates of correlation.

Maps also show a different kind of correlation, a *geographic correlation* distinct from the statistical correlation of the scatterplot and correlation coefficient. Statistical correlation is aspatial and reveals nothing about spatial trends. Figure 10.17 demonstrates this difference with two map pairs distinct in spatial pattern yet identical in scatterplot and correlation coefficient. Variables A and B, which share a comparatively chaotic, fragmented pattern, clearly differ in geographic correlation from variables X and Y, which have a distinct common trend with higher values toward the top of the region and lower values toward the bottom. Although not identical, the maps for X and Y suggest the influence of a third, underlying geographic factor, such as latitude, ethnicity, soil fertility, or proximity to a major source of pollution. Despite the problems posed by areal aggregation, the analyst of geographic data who explores correlation without also checking for spatial pattern is either ignorant, careless, or callous. And the nonskeptical reader is easily misled.

Places, Time, and Small Numbers

Areal data can yield particularly questionable patterns when choropleth maps show rates based on infrequent events, such as deaths from a rare type of cancer. Yet disease maps based on small numbers are a common tool of the epidemiologist, who uses mapping to explore the possible effects on human health of radon-rich soils, incinerators, and chemical waste dumps. But one question arises whenever the map shows a trend or cluster: Is the pattern real?

The problem is one of small numbers. Pandemics are rare, and seldom is the association between disease and an environmental cause so overwhelming that the link is easily identified and unchallenged. Clusters of deaths or diagnosed cases usually are few and unspectacularly small, perhaps no more than three deaths in a town or two in the same neighborhood. Epidemiologists map these cases both as points, to get a sense of patterning, and by areal units, to adjust for spatial differences in the number of people at risk. After all, an area with half the region's cases is not remarkable if it has half the region's population. But what is the significance of a small area with two or three cases and a rate several times above the national or regional rate? Could this pattern have arisen by chance? Would one or two fewer cases make the area no longer a "hot spot"? If one more case were to occur elsewhere, would this other area also have a high rate? To what extent does the pattern of high rates reflect arbitrary boundaries, drawn in the last century to promote efficient government or thirty years ago to expedite delivery of mail? Might another partitioning of the region yield a markedly different pattern? Might another level of aggregation—larger units or smaller units—alter the pattern? Is the mapping method inflating the significance of some clusters? And is it possibly hiding others?

Consider, for example, the maps in figure 10.18. At the top is John Snow's famous map showing cholera deaths clustered around the Broad Street Pump. A physician working in London during the cholera epidemic of 1854, Snow suspected drinking water as the source of infection. At that time homes

Snow's Dot Map

Areal Aggregations and Density Symbols

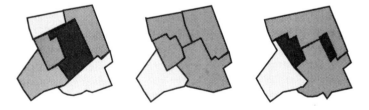

FIGURE 10.18. A reconstruction of John Snow's famous dot map of cholera (above) and three choropleth maps (below) produced by different areal aggregations of this part of London.

did not have running water, and people carried buckets from a nearby pump. Snow's map provided important evidence for the waterborne transmission of cholera; when authorities removed the pump's handle, new cases in this part of the city plummeted.

But what might have happened had Snow not worked with point data? The three maps at the bottom of figure 10.18 show how various schemes of areal aggregation might have diluted the Broad Street cluster. If addresses are available, as on most death certificates, aggregation to census tracts or other areal units larger than the city block increases the risk of missing intense, highly local clusters.

Aggregation involves not only areal units but also time, disease classification, and demography. One solution to the question of significance is to get more data by collecting information over a longer time span. Adding together several years of data, or even several decades, dampens the effect of chance occurrences but risks involving a wider range of causal agents. Aggregation over time might, for instance, mask important temporal trends, dilute the impact of new or abated environmental contaminants, or incorporate difficult-to-measure effects of population mobility. Likewise, combining several disease categories or the mortality of diverse demographic groups promotes stability and significance by increasing the number of cases and broadening the set of causes.

Clearly one map is not sufficient, although one good map can signal the need for a more detailed investigation. It is then up to a variety of scientific researchers to explore further the effects of geography and environment by examining employment and residential histories, characteristics of residence and neighborhood, and hereditary factors; by carefully studying maps at various levels of spatial, temporal, and demographic aggregation; through computer simulation to test the stability of known clusters and automated pattern recognition to identify new ones; and through related clinical and laboratory studies. Although maps can indeed lie, they can also hold vital clues for the medical detective.

Indexes, Rates, and Rates of Change

Another danger of one-map solutions is a set of measurements that presents an unduly positive or negative view. Often the map author has a single theme in mind and has several variables to choose from. Usually some variables are markedly more optimistic in tone or pattern than others, and the name of the index can cast a favorable or an unfavorable impression in the map title. "Labor Force Participation," for instance, sounds optimistic, whereas "Job Losses" clearly is a pessimist's term. An appropriately brazen title offers a good way to overstate economic health or industrial illness.

If the picture is bleaker or brighter than suits your politics, try a rate of change rather than a mere rate. After all, minor downturns often interrupt a run of good years, and depressions do not last forever. If unemployment is high now but a bit lower than a year, six months, or a month ago, the optimist in power would want a map showing a significant number of areas with declining unemployment. Conversely, the pessimist who is out of power will want a map depicting conditions at least as bad as before the current scoundrels took over. A time interval that begins when proportionately fewer people were out of work will make the opposition party's point, especially if unemployment has become worse in large, visually prominent, mostly rural regions.

A useful index for the optimist is one with relatively low values, such as the unemployment rate, if conditions have improved, or an index with comparatively high values, such as employment level, if conditions are worse. Thus a drop of one percentage point from a base of 4 percent unemployment yields an impressive 25 percent improvement! Yet a substantial increase in the unemployment rate from 4 to 6 percent can be viewed more optimistically as a drop in labor force participation from 96 to 94 percent—a mere 2 percent drop in employment.

Point symbols and counts, rather than rates, can be useful too. If the economy has been improving in all regions, the current government might want a map with graduated circles or bars showing actual counts beneath the title "Employment Gains." If the country is in a widespread recession, the opposition would use similar point symbols with the title "New Job Losses."

The cartographic propagandist is also sensitive to spatial patterns. Favorable symbols should be large and prominent, and unfavorable ones small and indistinct. Thus the optimist might present the unemployment data in figure 10.19 with the map at the lower left, to focus attention on improved conditions in larger areas, whereas the pessimist would prefer the map at the lower right, to emphasize the much greater number

Area	Labor Force (000s)	Unemployment (000s)			Unemployment Rate			Percentage Change
		t_1	t_2	Change	t_1	t_2	Change	
1	3,000	120	180	+60	4.0%	6.0%	+2.0%	+50.0%
2	16,000	640	800	+160	4.0%	5.0%	+1.0%	+25.0%
3	2,500	125	113	-12	5.0%	4.5%	-0.5%	-9.6%
4	800	56	48	-8	7.0%	6.0%	-1.0%	-14.3%
5	500	40	35	-5	8.0%	7.0%	-1.0%	-12.5%

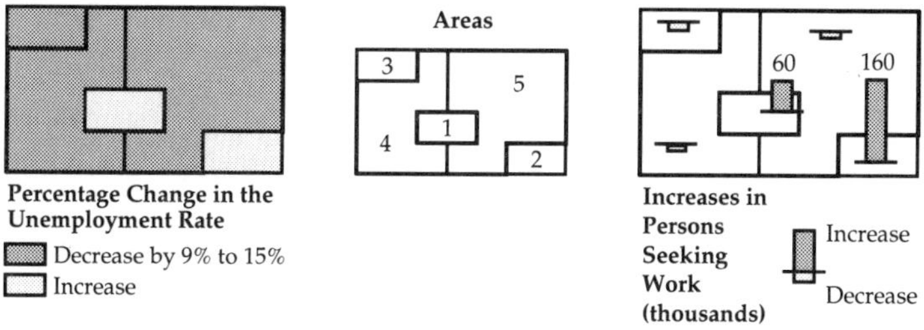

Percentage Change in the Unemployment Rate
▨ Decrease by 9% to 15%
☐ Increase

Areas

Increases in Persons Seeking Work (thousands)
▮ Increase
Decrease

FIGURE 10.19. Unemployment data (top) for a hypothetical region (bottom center) yield different maps, supporting an optimistic (bottom left) and a pessimistic (bottom right) view of recent temporal trends.

of unemployed persons in more urban areas. Note as well how the titles and keys in these examples reinforce cartographic manipulation.

Labor economists, who commonly adjust unemployment data for seasonal effects, discourage some manipulation of time intervals. After all, more people are seeking work in early summer, when many high-school and college graduates enter the labor force for the first time. And more people find at least temporary work in November and December, the peak shopping season. Local seasonal effects, such as tourism and the temporary hiring of cannery workers in agricultural areas, also require seasonal adjustment.

Mortality, fertility, and other phenomena that do not affect all segments of the population equally also require adjustment. Figure 10.20, a comparison of the age-adjusted death rate with the crude death rate, illustrates the advantage of mapping demographically adjusted rates. The map at the left is a simple rate, which does not consider such age differences as a relatively young population in Alaska and older populations in Arkansas and Maine. When the rates portrayed in the right-hand map are adjusted for age differences, Alaska and some southeastern states emerge as high-rate areas whereas the

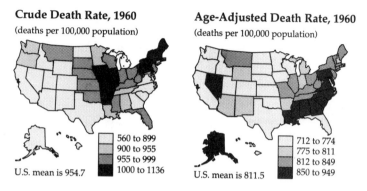

Crude Death Rate, 1960
(deaths per 100,000 population)

560 to 899
900 to 955
955 to 999
1000 to 1136

U.S. mean is 954.7

Age-Adjusted Death Rate, 1960
(deaths per 100,000 population)

712 to 774
775 to 811
812 to 849
850 to 949

U.S. mean is 811.5

FIGURE 10.20. Maps of the crude death rate (left) and the age-adjusted death rate (right) can present markedly different geographic patterns of mortality.

Northeast and Midwest slip to a lower category. Age-adjustment allows the map at the right to reveal the effects of relatively good health care and a higher socioeconomic status in the New England, Middle Atlantic, and North Central states, widespread poverty and less accessible health care in the South of the 1960s, and the effects of accidents and isolation in Alaska.

When a single variable might yield many different maps, which one is right? Or is this the key question? Should there be just one map? Should not the viewer be given several maps, or perhaps the opportunity to experiment with symbolization through a computer workstation? If unable to trust the presenter's honesty and thoroughness, the skeptical viewer must question the representativeness of a single graphic. Guard against not only the cartographic manipulator, but also the careless map author unaware of the effects of aggregation and classification.

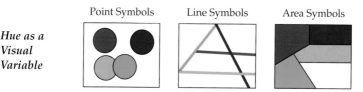

Point Symbols   Line Symbols   Area Symbols

PLATE 1. For area symbols in particular, hue is often more forceful than the other five principal visual variables (fig. 2.11).

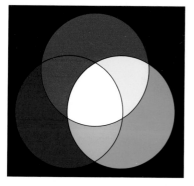

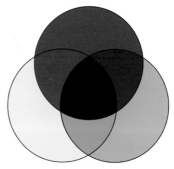

Additive Primary Colors   **Subtractive Primary Colors**

PLATE 2. Primary colors yield other hues when combined as either beams of light (left) or patches of dye (right).

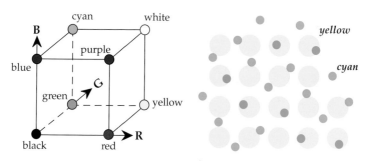

RGB Color Cube   **Screened Process Color: Green**

PLATE 3. RGB color cube (left) and greatly enlarged representation of over-printed screens of colored dots used in process printing to produce green (right).

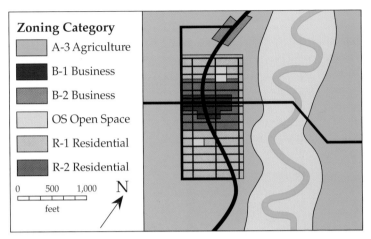

PLATE 4. Contrasting hues efficiently describe qualitative differences on zoning map shown in monochrome in figure 6.1.

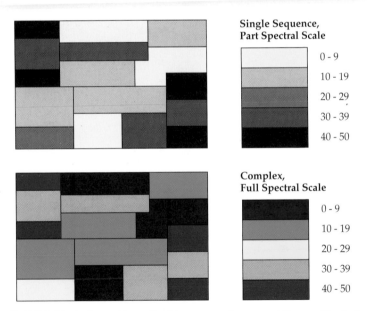

PLATE 5. Limited set of hues (top) is more easily grasped than an illogical, complex sequence of spectral hues (bottom).

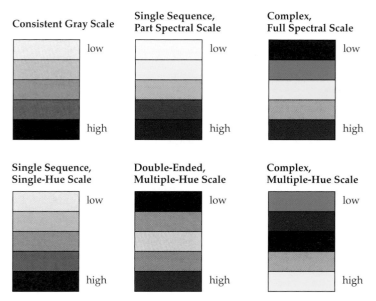

Consistent Gray Scale

low

high

Single Sequence, Part Spectral Scale

low

high

Complex, Full Spectral Scale

low

high

Single Sequence, Single-Hue Scale

low

high

Double-Ended, Multiple-Hue Scale

low

high

Complex, Multiple-Hue Scale

low

high

PLATE 6. Some color sequences found on choropleth maps.

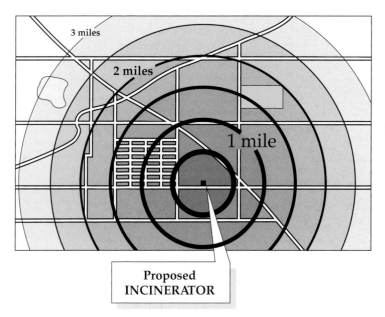

PLATE 7. Red area symbols connoting increased danger near the site of a proposed incinerator strengthen the message of a monochrome environmental propaganda map (fig. 7.19).

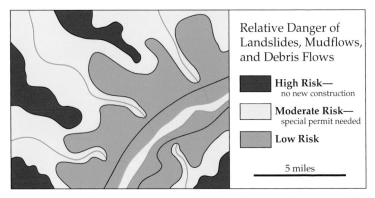

PLATE 8. The map viewer reminded of the graphic metaphor can readily decode a sequence of three traffic-light colors portraying degree of environmental risk.

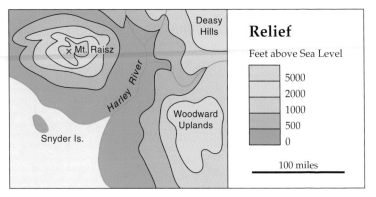

PLATE 9. Although hypsometric tints are widely used to portray relief with color-coded elevation categories, map viewers must be aware that lowland areas shown in green might well be dry and barren.

PLATE 10. Squares at the center of these three boxes are identical, but because of simultaneous contrast the gray center appears darker when surrounded by a more brilliant color.

Chapter 11

COLOR
ATTRACTION AND DISTRACTION

☀

Color is a cartographic quagmire. Color symbols can make a map visually attractive as well as fulfill the need for contrast on road maps, geological maps, and other maps with many categories. Yet the complexity and seductiveness of color overwhelm many mapmakers, and countless maps in computer graphics demonstrations, business presentations, and daily newspapers reveal a widespread ignorance of how color can help or hurt a map. Persons unaware of the appropriate use of color in cartography are easily impressed and might accept as useful a poor map that merely looks pretty.

Technological change accounts for much of the misuse of color on maps. Before the 1980s color printing was expensive and seldom used thoughtlessly, and color maps were comparatively rare. Advances since 1980 in electronic computing and graphic arts have encouraged a fuller use—and abuse—of color. Inexpensive color monitors, color printers, and slide generators have made color effortlessly available to the amateur mapmaker, and run-of-press color lithography encourages a similar misuse by cartographically illiterate commercial artists, responsible for most news illustration. Moreover, many viewers and readers expect maps with richly contrasting hues, even when black-and-white or more subdued symbols might be more readily and reliably decoded. This chapter briefly explains the nature of color and examines how graphic logic, visual perception, and cultural preferences affect the use of color on maps.

The Phenomenon of Color

As a biophysical phenomenon, color is a sensory response to electromagnetic radiation in a narrow part of the wavelength spectrum between roughly 0.4 µm and 0.7 µm, called the "visible band." (One micrometer [µm] is one-millionth of a meter.) The eye cannot see shorter-wave radiation, such as ultraviolet light (10^{-1} µm) or gamma rays (10^{-6} µm), nor can it sense longer-wave energy, such as microwave radiation (10^5 µm) and television signals (10^8 µm). Yet within the visible band the eye and brain readily distinguish among wavelengths associated with the hues we call violet, blue, green, yellow, orange, and red, as in the left half of figure 11.1. White light is a mixture of all these wavelengths, but a rainbow or prism can refract white light and sort its constituent colors into this familiar spectral sequence.

As a perceptual and graphic-arts phenomenon, color has three dimensions: hue, value, and saturation. *Hue,* related to the wavelength of electromagnetic radiation, is the most obvious gauge; when most people think of color, they think of hue (plate 1). *Value,* which refers to a color's lightness or darkness, applies to both hues and shades of gray. *Saturation,* also called *chroma,* refers to a color's intensity or brilliance; a medium blue, for instance, might range in saturation from a pure, 100 percent saturated strong blue down through moderate blue, weak blue, and bluish gray, to a 0 percent saturated medium gray. Graytones, also called *achromatic color,* have zero saturation.

Color theorists often invoke the HVS color-space diagram shown in the right half of figure 11.1 to describe the three-

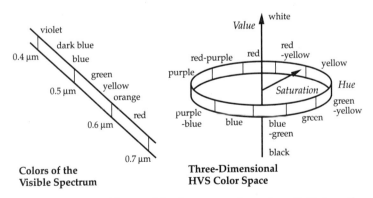

Colors of the Visible Spectrum

Three-Dimensional HVS Color Space

FIGURE 11.1. The visible part of the electromagnetic spectrum (left) shows the relation between wavelength and hue. The HVS color space (right) links hue with the other two dimensions of color: value and saturation.

dimensional relationship of hue, value, and saturation. A scale similar to the visible band but bent into a circle that joins dark red with dark purple represents the range of hues. This circle of hues, or *color wheel,* is centered on and perpendicular to a vertical axis of graytone values, which range from black at the bottom to white at the top. Saturation, the third dimension, measures a color's distance outward from this graytone axis toward one of the pure hues on the color circle; a fully saturated blue, for instance, would lie farther from the value axis than a weak blue.

As visual variables, hue and value play very different roles. Although we perceive blue, green, and red as somehow different, we readily organize various shades of red along a scale from light red to dark red. Designers cannot manipulate chroma as easily as hue and value, and they seldom consciously use it as a visual variable.

Colors on maps usually are mixtures of primary colors. CRT (cathode ray tube) display monitors generate a wide variety of hues by mixing the *additive primary colors* of red, green, and blue. The left half of plate 2 demonstrates some of the mixed colors occurring when red, green, and blue spotlights produce overlapping circles on a white, reflective stage. Yellow occurs when green light is added to red light, cyan is a hue between blue and green, and magenta is a reddish purple reflecting the addition of red and blue. Overlapping red, green, and blue circles on a color monitor or television screen would yield the same effect. Note that white occurs at the center, where all three circles overlap. White light results from mixing red, green, and blue wavelengths of roughly equal intensity. Outside the area covered by one or more spotlights, the stage appears black, indicating the absence of color.

Recognition of white light as the summation of red, green, and blue makes it easy to understand the *subtractive primary colors.* Yellow, magenta, and cyan are called subtractive primaries because each represents the subtraction of one of the additive primary colors from white light. For example, a white card dyed yellow will absorb blue light from white light and reflect the remaining mixture of red and green, which the eye perceives as yellow. That is, white (which is red + green + blue) minus blue (removed by the yellow dye) leaves red plus green, which is yellow. Similarly, a magenta dye removes green wavelengths from a white-light mixture, and a cyan dye removes red light from white light. Printed maps have color because dyes on the paper are selectively absorbing components of white light, whereas maps on color monitors and color TV screens have color because phosphor dots are emitting electromagnetic radiation in the red, green, or blue parts of the spectrum.

As with the additive primary colors, mixing appropriate amounts of yellow, magenta, and cyan can generate other hues. The right half of plate 2 shows the pattern of colors produced by overlapping circles dyed with the three subtractive primaries. Note that an additive primary color occurs where only two circles overlap and that black (the absence of color) occurs where all three overlap. Red, for instance, occurs where yellow and magenta dyes overlap because subtracting blue and green respectively from white light leaves only red. Black occurs in the triple-overlap center zone because the three dyes together subtract all of the additive primaries.

Primary colors simplify both electronic display and printing. Colors on CRT displays are based upon the additive primaries and three separate arrays of thin, closely spaced phosphor dots or lines, 50 or more to the inch and each addressed by a separate electron gun. (If this concept is unfamiliar, look very closely at a color TV or color monitor—but not for long!) In an area of the screen, each set of intermingled red, green, and blue dots or lines can be excited to emit light at a different intensity. If 64 different intensity levels are provided for each color, the RGB (for Red-Green-Blue) color monitor can in theory produce 262,144 (64^3) different colors. A three-dimensional diagram called the RGB color cube, as in the left half of plate 3, represents the range of color possible with an RGB color monitor; any point within the cube represents a unique color, defined by unique coordinates for the cube's three axes. In practice, graphics systems often restrict the user to a *palette* with a much smaller, more manageable number of discrete choices.

Sets of intermingled dots are also the basis for printed colors, but these "screened" dots vary in size rather than in intensity and absorb rather than emit light. A technique called *process printing* depends upon a reflective, light background and uses transparent inks of the three subtractive primary hues to absorb varying amounts of red, green, and blue. Thus an area with comparatively large yellow dots and small cyan dots printed on white paper, as in the right half of plate 3, will absorb a moderate amount of blue light as well as some red light and will appear green. Because the dots are small and closely spaced—a typical screen arranges the dots in a fine grid with rows of 50 to 150 tiny dots *per inch*—our eyes ignore the screens' texture and orientation but mix their dyes. In process printing, a fourth impression is made using black ink to assure a strong black for type and fine lines.

Maps televised in color often defy the designer's intent. Television sets might be poorly tuned or otherwise defective, or the screen might reflect glare from a bright light or the sun. Because reception or viewing conditions can destroy subtle color differences for large numbers of viewers, televised maps tend to have highly contrasting hues, even when the data represent differences in intensity rather than kind. Moreover, because picture-tube manufacturers have not fully standardized their products, two screens might display noticeably different versions of the same color. Printing can also thwart the map author's intent, especially where the printing press transfers extra ink to the paper.

Color on Maps

How can you tell when a map in color might be misleading? The wary map user must first ask whether the map uses color—that is, hue—to portray differences in intensity or differences in kind. Soils maps, geologic maps, climatic maps, vegetation maps, zoning maps (see fig 6.1 and plate 4), land-use maps, road maps, and many other types of maps showing a variety of features can benefit from contrasting hues, provided that somewhat similar hues represent somewhat similar features and radically different hues represent radically different features. Be suspicious, though, when contrasting hues attempt to show differences in intensity on choropleth maps (discussed in chap. 10).

Hue differences usually fail at portraying differences in percentages, rates, median values, and other intensity measures because spectral hues have no logical ordering in the mind's eye. Consider the experiment described in figure 11.2, a mass

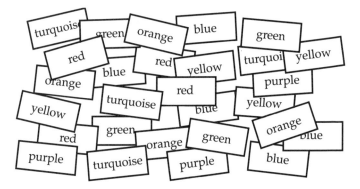

FIGURE 11.2. People tend not to be able to sort a variety of spectral hues into a single consistent sequence of colors.

of cards in seven different hues, all with the same value and chroma. If ten people were asked separately and without collaboration to choose seven cards of different colors and arrange them in order from low to high, the odds are they would produce ten different sequences. Some might order them from green to red, some from blue to red, and some might attempt to reproduce a rainbow sequence. Two people might even agree on the same series but disagree on its direction from low to high. The experiment would demonstrate clearly that there is no simple, readily remembered and easily used sequence of hues that would obviate a map reader's need to refer back and forth repeatedly between map and key. Color symbols can be used on choropleth maps, but not conveniently. The use of spectral hues to portray intensity differences is a strong clue that the mapmaker either knows little about map design or cares little about the map user.

Not all use of color for choropleth maps is confusing or undesirable. Value differences and hues coexist nicely in some single-sequence, part-spectral color scales, such as the sequence from light yellow to black in the upper map in plate 5. Note that this progression from light yellow to brown-yellow to brown to dark brown to black has a consistent, logical, and readily comprehended ordering from light to dark. Not only does this sequence of yellows, browns, and black show the pattern of high and low values as effectively as the gray-tones in figure 11.3, but the color map is more aesthetically appealing. In contrast, although young children and some adults might be attracted to the lower map in plate 5, most users will find its full spectral scale of primary hues confusing, complex, and comparatively difficult to decode.

Plate 6 demonstrates the variety of color scales found on choropleth maps. The gray scale in the upper left might, for example, adopt a single hue, as at the lower left, without any loss in consistency, graphic logic, or ease of use. A partial spectral scale based on yellow, orange, and red, as in the middle of the top row, can be as consistent, convenient, and visually appealing as the yellow–brown–black sequence in plate 5; yellow–green–dark blue provides another common single-sequence, part-spectral color scale useful on choropleth maps. Among the readers of color weather maps, the more graphically complex spectral scale at the upper right not only draws upon useful associations of blue with cold and red with hot, but also upon the reinforcement of daily exposure to the same color scheme; yet for most choropleth maps this full-spectral sequence is almost as awkward and troublesome a way to introduce color as the multihue, nonspectral sequence at the lower right. In comparison, the double-ended scale in the middle of the lower row uses contrasting hues to support a readily comprehended graphic logic, useful for maps showing both positive and negative rates of change. In this scheme, the colder blues represent significant losses or declines, the warmer reds indicate major gains or increases, neutral grays represent minor change, and weak blues and reds show moderate decreases and increases, respectively. When reinforced on

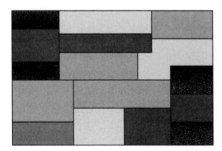

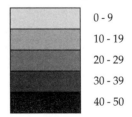

Consistent Gray Scale

	0 - 9
	10 - 19
	20 - 29
	30 - 39
	40 - 50

FIGURE 11.3. A logical, consistent sequence of graytones describes intensity variations more reliably than a complex, graphically illogical sequence of spectral hues (plate 5, bottom).

a succession of maps, this color scale can be a powerful tool for mapping historical economic data.

Because colors distinctly different in hue can be highly similar in value, map users should be suspicious of maps received by fax or reproduced on a black-and-white photocopier. Dark red and dark blue can copy as black or dark gray, for instance, while yellow might turn white. Map authors too should be mindful of the "Xerox effect," and anticipate electronic distortion of maps with complex colors. In the spirit of government reports with "intentionally left blank" stamped on otherwise empty pages, maps might include a small note advising whether they were printed in color or monochrome.

The conscientious map user must also be wary of the deliberate or inadvertent use of color to make a feature or proposal appear attractive or unattractive. People respond emotionally to some colors, such as blue and red, and some of these responses are common and predictable enough to be tools of the cartographic propagandist. And even if no deliberate manipulation is intended, because of embedded emotions or culturally conditioned attitudes some colors carry subtle added meaning that could affect our interpretation of a map or our feelings about the map or the elements it portrays.

Color preferences vary with culture, life cycle, and other demographic characteristics. For instance, men tend to prefer orange to yellow and blue to red, whereas women favor red over blue and yellow over orange. Preschool children like highly saturated colors, such as bright red, green, and blue, whereas affluent middle-aged adults generally prefer more subtle, pastel shades. Among the spectral colors, North American adults seem to prefer blue and red to green and violet, and green and violet to orange and yellow. Least appreciated is a vomitlike greenish yellow.

Preferences also extend to groups or ranges of colors. A range of greens and blues, for instance, is generally preferable to a range of yellows and yellow-greens. Among people who like earthtones, a yellow–brown sequence would be attractive.

Little is known about the effects upon map users of a variety of subjective reactions to color. Most colors, in fact, relate to several concepts, favorable and unfavorable. Red, for instance, is associated with fire, warning, heat, blood, anger, courage, power, love, material force, and Communism, and its effect probably depends very much upon context. A right-wing political group might conveniently paint the Former Soviet Union, Cuba, or China red on a world map, for instance, whereas a marketing manager might use a map with red target areas to focus the attention of corporate managers. Plate 7 illustrates how the addition of progressively redder, more intense tints makes a forceful propaganda map (fig. 7.19) even stronger. Similarly, black might connote mourning, death, or heaviness, whereas blue can suggest coldness, depression, aristocracy, or submissive faith. White might suggest cleanliness or sickness, and green can relate to envy, compassion, or the Irish. Yellow's subjective message, if any, clearly depends on context: its possible use as a symbol of weakness and cowardice contrasts with its almost equally strong association with cheerfulness and power.

Some designers find color a clever or an obvious reinforcement to pictorial symbols. Given a golden tinge and used over a dollar sign, for instance, yellow reinforces an icon of wealth. Other examples of color-enhanced icons are green shamrocks, symmetrical red crosses on ambulances or hospitals, vertical black crosses for cemeteries or churches, and lemon yellow cars with flat tires.

Maps portraying environmental hazards often borrow the familiar red–yellow–green sequence of traffic light hues, as in the landslide-hazard map in plate 8. This sequence is highly effective, at least among map viewers who drive, because of continually reinforced associations of red with danger, yellow with a need for caution, and green with lower risk. Even so, reliable use of these colors requires the map author to explain the metaphor, perhaps with a stop-light icon in the map key.

Color's effectiveness as a map symbol might conflict with or reinforce its role as a landscape metaphor. For several centuries, cartographers have exploited and encouraged such associations as green with vegetation, blue with water, red with high temperatures, and yellow with a desert environment, and where the context is correct and appropriate, these associations promote efficient decoding. But some caution is warranted, for the blueness of the water might exist largely in the minds of wishful environmentalists, self-serving tourist operators, and gullible map readers.

A cartographic scheme called *hypsometric tints* can be partic-

ularly misleading to map users unaware of its focus on elevation. Plate 9 illustrates a common practice for portraying relief
on general-purpose wall maps and atlas maps: a multihue
color scale with five or more steps ranging from medium green
for low elevations through yellow to orange or white for the
highest elevations. Although widely used, hypsometric tints
have not been fully standardized, and a variety of variations
and modifications occur, including dark green for elevations
below sea level and a dark, rusty brown for higher elevations.
Confusion is likely if the reader associates white with snow,
green with abundant vegetation, and yellow or brown with
desert—much of the world's tundra is close to sea level, many
lowland areas are deserts, and many upland areas have substantial forests or grasslands.

Conscientious users of color maps must also be wary of
simultaneous contrast, the eye's tendency to perceive a higher
degree of contrast for juxtaposed colors. When a light color is
surrounded by a dark color, as in plate 10, simultaneous contrast will make the light color seem lighter and the dark color
seem darker. Thus a medium gray or blue surrounded by
darker symbols will appear lighter, whereas its sample in the
map key is surrounded by white and appears darker. This
effect can be particularly troublesome on geological maps and
on some environmental maps with many categories, especially
when only slightly different colored symbols represent
markedly different categories.

Another perceptual effect that causes confusion between
colors on the map and colors in the key is the tendency for
large patches of color to look more saturated than small
patches of the same color. For example, a large area of moderate green might appear to match a small sample of bright
green in the key. On maps with many categories and varied
colors, the good cartographer provides a redundant stimulus,
such as alphanumeric codes or patterned area symbols printed
in black, to help the conscientious map reader use the map key
accurately.

Map viewers searching a color map for a place or feature
must watch for inadvertent camouflage. Because geography
can juxtapose odd colors, poor planning by the map author
often leads to poor contrast between type and point symbols
and their background, including such atrocities as yellow type

on a white background and purple or blue type on a black background. Both combinations produce labels that are easily overlooked and difficult to read. Yet yellow works well against black, and purple can be highly legible on white, especially if lighting is poor. Type is likely to be illegible when a label must cross both light and dark backgrounds.

Color on televised maps is particularly adept at serving two masters poorly. Providing value contrast for viewers with black-and-white sets limits the effective use of contrasting hue on TV maps. Yet not designing for black-and-white compatibility makes the black-and-white television viewer as vulnerable as a person with color-deficient vision. Among the worst combinations is red type on a green background, where poor contrast in value easily hides a label on a black-and-white screen.

Personal computers are another source of bad maps in color. Lacking experience with electronic displays and additive colors, amateur mapmakers often mimic printed maps on a computer. Yet color monitors have dark backgrounds instead of the more familiar white, and video graphics with large amounts of white can "bloom" and irritate the eye. Moreover, color palettes can severely limit the colors and grays available and thus force the use of color by precluding an ordered range of graytones. The problem need not lie with the computer—programmers with no training in cartography and little sense of graphic design have been highly successful in writing and marketing mapping software. With no guidance and poorly chosen standard symbols, users of mapping software are as accident-prone as inexperienced hunters with hair-trigger firearms. If you see one coming, look out!

Chapter 12

MULTIMEDIA, EXPERIENTIAL
MAPS, AND GRAPHIC SCRIPTS

Computer technology is undermining the two most convincing excuses for little white cartographic lies. Although cost of production and competition for space account for the current dearth of complimentary maps portraying the same distribution, these constraints are hardly relevant to interactive, *experiential* maps, which users can explore freely on a home computer. In addition to liberating map viewers from static single-map representations, multimedia not only affords a more dramatic and informative cartographic treatment of dynamic phenomena such as wars and explorations but also promotes the integration of maps, graphs, pictures, written text, and sound. Besides encouraging map authors to develop narrative presentations, the new technology supports customized guided tours of electronic databases, tailored to the individual user's interests and background. Yet despite this emancipation from rigid paper maps, electronic maps are no less prone to bias and exploitation.

This chapter explores the advantages and pitfalls of dynamic cartography, in which experiential maps, often richly informative, can prove seductive or frustrating. The first section examines the interactive analysis of geographic information and warns of bias in both software and data. The second section addresses graphic narratives, which (like essays or films) depend on linear coherence and are prone to bias and distortion. Despite these threats, multimedia fosters informative tutorials as well as graphic narratives that evaluate data quality and screen databases for interesting, potentially meaningful geographic patterns.

Of Mice and Menus

Although the term *direct manipulation* might not be familiar, most readers will have seen, if not used, a computer with a *menu bar* across the top of the screen and a *mouse* or other pointing device for selecting and moving words and graphic objects. Introduced in the early 1980s, the direct-manipulation interface lets the map viewer control much of the content and form of an electronic map—snapshots of which can be saved for future reference or used to illustrate a report or talk. Although the paper map is far from dead, computer graphics has radically altered how we look at and explore geographic information.

A hypothetical software application illustrates several of interactive cartography's more liberating innovations. Known as SECSS, for Social Engineering Cartographic Support System—enticing acronyms are *de rigueur* in the software business—our trusty mapping system greets the viewer with a menu bar offering four pull-down menus. As figure 12.1 illustrates, the viewer who moves the pointer to the word "Variable" and presses a button on the mouse may then select a map theme from a list of data. A check mark indicates that the viewer has elected to map unemployment rates. Using the

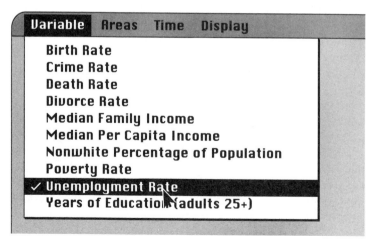

FIGURE 12.1. Snapshot of the upper-left corner of the screen shows the variety of map themes offered in a pull-down menu selected from the menu bar.

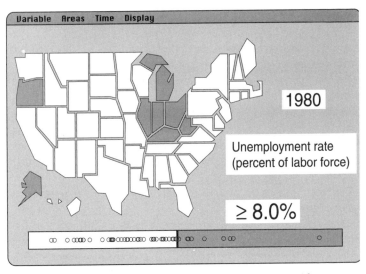

FIGURE 12.2. Interactive number line lets viewer experiment with two-category choropleth map.

other pull-down menus, our user then selects the 50 United States as the area, 1980 as the time, and an interactive two-class choropleth map as the display. Figure 12.2 shows the results of these selections: an interactive number line on which a movable break divides the data into two classes, and a state-level map on which dark shading highlights states in the higher category. Using the mouse, the viewer explores the data by shifting the category break left or right along the number line. For example, moving the break far to the right, as in figure 12.3, reveals that Michigan, with a rate above 12 percent, led the nation in joblessness.

To obviate a tedious year-by-year examination of historical trends, SECSS provides a sliding marker, or *slider* (fig. 12.4), controlled by the mouse, for moving back and forth in time. To obtain this view, the user selected a three-category temporal map from the display menu, and the system responded initially with the time-series graph in figure 12.5, showing individual trend lines for each state and providing sliders for two category breaks. (Because the display contains more than one slider, the viewer must point to each category break separately and hold the mouse button down while "dragging" its slider to a new position.) The temporal map in figure 12.4 requires

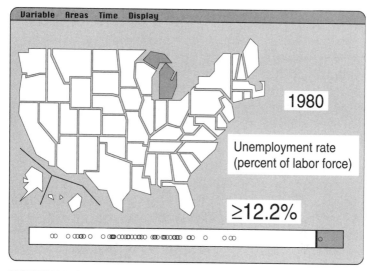

FIGURE 12.3. Category break near right end of number line reveals state with highest unemployment rate.

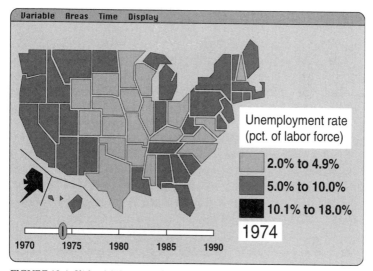

FIGURE 12.4. Slider (sliding marker) below map lets viewer explore temporal change in the unemployment rate.

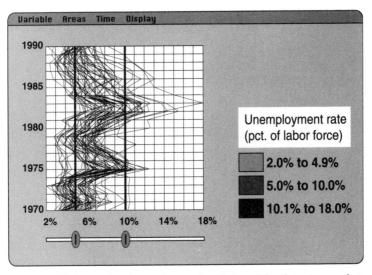

FIGURE 12.5. Interactive time-series graph with two sliders lets viewer select both category breaks for the three-class temporal map in figure 12.4.

fixed class breaks, and the graph helps the viewer avoid categories that change either too seldom or too often as the viewer moves back and forth in time.

The map and graph juxtaposed in figure 12.6 describe another asset of experiential cartography: integrative visual analysis of spatial-temporal data through *linked windows.* Because each trend line on the graph is linked to a polygon on the map, highlighting a state also highlights the state's temporal trend. Similarly, selecting a trend line on the graph simultaneously highlights the temporal trend's location on the map. In this example, the viewer has linked the graph's most prominent spike, a joblessness rate of 18 percent in 1983, to West Virginia. The viewer curious about specific regions or similar temporal patterns can hold down the mouse button and select several areas or trend lines at once. Linked windows are also useful in exploring bivariate relationships visualized by juxtaposing a map and a bivariate scatterplot (fig. 10.17).

Direct manipulation supports a wider variety of cartographic tasks. Electronic systems for viewing topographic maps provide several options for changing scale. The viewer can elect a larger-scale, more detailed view by drawing a rec-

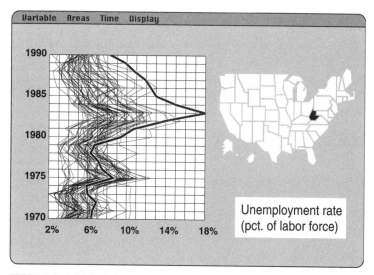

FIGURE 12.6. Linked objects can be highlighted simultaneously in both map and graph.

tangle on a portion of the screen, or by marking a focal point on the map and then zooming in through a series of progressively more detailed views. To find a specific place, the viewer can scroll around the map in two dimensions or type in the place's name or postal code. Because multimedia supports sound, an electronic atlas might even enunciate the preferred local pronunciation of place-names fingered with the mouse. Especially useful is the interactive query function that displays additional information about places or features pointed to on the map—a handy tool for finding fast-food restaurants at freeway interchanges or retrieving supplementary facts about anomalous areas on a data map. More advanced systems offer perspective or three-dimensional views of terrain, including dramatic fly-by sequences and virtual-reality walkovers, in which the viewer who dons a helmet and mounts a treadmill can simulate the sights and sounds of being there.

However flexible and interactive the system, the user can see only what the software and data allow. For example, if SECSS provides only annual data aggregated by state, its maps will obscure significant seasonal variations as well as noteworthy local patterns. Rigid data are especially frustrating to the policy maker eager to disaggregate unemployment rates by gender, race, age, educational attainment, or occupation. Even so, a rich database is not necessarily a reliable database. Most geographic data systems, sad to say, offer no assessment of data quality, and many don't even describe underlying measurements. Are you aware, for instance, that the unemployment rate can fall—quite artificially—when conditions become so bad that people who want a job become discouraged and stop looking? By seducing viewers into believing that the data are reliable, relevant, and essentially complete, a geographic information system can become a dangerous instrument of self-deception.

Graphic Narratives

Graphic narratives carry the yin-yang of insight and deception a step further by providing a focused sequence of maps, text, graphs, diagrams, and other images addressing a specific communication goal. A particularly useful narrative is the guided tour that introduces the viewer to new software or a new database. Multimedia atlases are ripe for carefully authored graphic narratives addressing a variety of topics—continental drift, the colonial slave trade in Africa and North America, the Revolutionary War, or the breakup of Yugoslavia, to name a few. Because viewers cannot easily read text and examine images at the same time, a verbal commentary can improve comprehension by providing an interpretation as well as pointing out what viewers should look for.

Although many graphic narratives are little more than slide shows recorded on a computer, generalized sequences of views called graphic scripts can be powerful tools for exploring geographic data. As a simple example, figure 12.7 illustrates how a graphic script can expedite an otherwise tedious task, namely, the repeated display of maps showing rates of change for various time intervals. This particular snapshot is one in a sequence of 16 maps, each portraying a five-year rate of change for the period 1970 through 1990. The first map in the sequence covers 1970–75, the second 1971–76, and so forth. A dynamic time scale below the map represents graphically the duration of the time period as well as the starting and ending years for each individual view; as the narrative proceeds, the black bar moves from left to right across the time scale. In

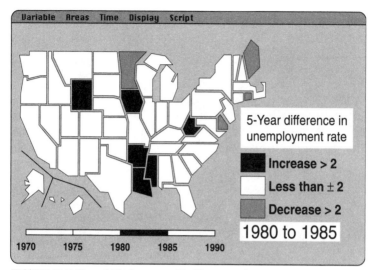

FIGURE 12.7. One of 16 views provided by a graphic script generating five-year rate-of-change maps from annual data for the period 1970 through 1990.

activating the script, the viewer need only specify the time interval, five years in this example. Another graphic script might generate rate-of-change maps for a variable time interval anchored to a single year. Anchoring the sequence at 1988, for instance, would yield rate-of-change maps for 1970–88, 1971–88, and so on. Scripts based on a common set of data, base maps, and class intervals are easy to set up in the programming languages used to develop mapping software.

Applying concepts in artificial intelligence and expert systems, sophisticated graphic scripts can orchestrate an exploration of geographic data and even evaluate data quality. Novice viewers can thereby draw on the experience and wisdom of experts in spatial analysis. Addition of a *user profile* describing the viewer's interests and analytical prowess enables the script to generate a sequence of graphics incorporating uniquely relevant categories and comparisons. In examining unemployment data, for instance, a graphic script might enlighten a long-time resident of New York with the dynamic temporal map in figure 12.8, which highlights regions with higher unemployment than the Empire State. A customized script might also describe intrastate patterns (on a county-unit map) for important sectors of the state's economy, adjust for

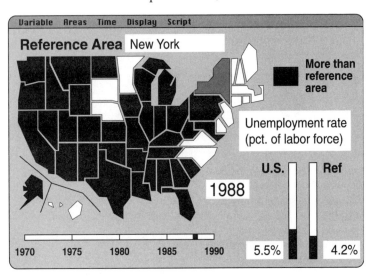

Variable   Areas   Time   Display   Script

Reference Area New York

More than reference area

Unemployment rate (pct. of labor force)

U.S.   **Ref**

1988

1970   1975   1980   1985   1990

5.5%   4.2%

FIGURE 12.8. One of 21 views in a dynamic temporal map produced by a graphic script tailored to the unique concerns of New York State residents.

seasonal effects and the national business cycle, and present graphs, tables, and text identifying New York's role in the world economy and assessing the effects of state and federal industrial policy. A thorough analysis would, no doubt, also question the reliability of the data and explore the vulnerability of mapped patterns to typical amounts of error.

A related concept is the automated prescreening of data for potentially intriguing, possibly meaningful geographic patterns. Letting the computer do the looking can be especially helpful to the epidemiologist searching for clusters of cancers and other diseases possibly affected by local air pollution or contaminated groundwater. Letting the computer run overnight with a prescreening script that tries out a vast variety of data classifications is a good strategy for focusing the researcher's attention on a manageably small number of incisive maps that merit careful examination.

By harnessing patterns of analysis employed by recognized experts, graphic scripts can promote vigilance and thoroughness as well as encourage insight. Even so, experts are not without bias, particularly the flamboyant ones eagerly promoting a vaguely defined model or index that in an earlier era might reek of snake oil. Skeptical viewing is especially prudent for narratives by authors pursuing (or likely to pursue) a particular political or disciplinary agenda. The wary viewer can, of course, compare graphic scripts from diverse experts in much the same way that the conscientious journalist seeks a balanced mix of opposing opinions.

Viewers should be suspicious of systems that present processed data as irrefutable facts and forbid interference with the narrative's progress. The conscientiously curious viewer needs to stop, think, back up, and probe the data interactively. Access to raw data is essential, for instance, in a scripted presentation of an environmental impact statement. After all, the parties affected by a proposed shopping mall, landfill, or incinerator should be able to assess the appropriateness of observations and measurements purported to assure the project's neighbors that everything will be all right.

Because maps can be strongly persuasive, the graphic narrative is a potent rhetorical weapon. Odds are, though, that at least some viewers—informed skeptics like you—will not be so readily convinced. As display systems become more flexible, and more like video games, users must be wary that maps, however realistic, are merely representations, vulnerable to bias in both what they show and what they ignore. Skepticism is especially warranted when a dynamic map supporting a simulation model pretends to describe the future. Although electronic cartography may make complex simulations easier to understand, no one should trust blindly a map that acts like a crystal ball.

Chapter 13

Epilogue

The preceding chapters have explored the wide variety of ways maps can lie: why maps usually must tell some white lies, how maps can be exploited to tell manipulative lies, and why maps often distort the truth when a well-intentioned map author fails to understand cartographic generalization and graphic principles. The wise map user is thus a skeptic, ever wary of confusing or misleading distortions conceived by ignorant or diabolical map authors.

Let me conclude with a cautionary note about the increased likelihood of cartographic distortion when a map must play the dual role of both informing and impressing its audience. Savvy map viewers must recognize that not all maps are intended solely to inform the viewer about location or geographic relationships. As visual stimuli, maps can look pretty, intriguing, or important. As graphic fashion statements, maps not only decorate but send subtle or subliminal messages about their authors, sponsors, or publishers. Some advertising maps, for instance, announce that a power company or chain restaurant is concerned about the city or region, whereas free street guides attest to the helpfulness of a real estate firm or bank. A flashy map, in color with an unconventional projection, touts its author's sense of innovation, and cartographic window dressing in a doctoral dissertation or academic journal suggests the work is scholarly or scientific. An ornate print of an eighteenth-century map of Sweden not only decorates a living room wall but proclaims the household's pride in its Scandinavian heritage. A world map behind a television newscaster reinforces the network's image of excellence in global news coverage, and a state highway map is a convenient vehicle for a political message from the governor, image-building photos of the state's tourist attractions, and a cartographic statement about tax dollars well spent on roads, recreation sites, and forest preserves. A local map titled "Risk of Rape" can shock and can advocate more diligent police patrols and stricter sentencing. A cartogram comparing wealth or life expectancy among the world's nations can foster complacent pride or evoke compassionate guilt.

Maps with dual roles are not inherently bad. Indeed, some perfectly correct maps exist primarily to lend an aura of truth, and others exist largely as visual decoration. The impetus for an increased use of news maps was the perception among publishers that a better "packaged," more graphic newspaper could compete effectively with television as well as with rival papers. Their motivation might not have been better reporting, but the conscious decision to use more maps has improved their coverage of many news stories in which location is important. Similarly, competition for audience attention has led to more news maps in the electronic media; local television stations offer highly informative sequences of weather maps, and network news programs usefully complement the newscaster's "talking head" with simple yet instructive maps of relative location for major news events. Maps intended to decorate or impress can educate a public appallingly ignorant about basic place-name geography. Were it not for the map's power as a symbol of geographic knowledge, we would know a great deal less about our neighborhoods, our nation, and the world.

Dual motives are risky, of course. Map authors pursuing aesthetic goals might violate cartographic principles or suppress important but artistically inconvenient information. Maps, like buildings, suffer when the designer puts form ahead of function. Map authors with propagandist motives might suppress ideologically inconvenient information as well as knowingly adopt an inappropriate projection or dysfunctional symbols. And expedient map authors distracted by a need to decorate can deliver sloppy, misleading maps. The skeptical map viewer will assess the map author's motives and ask how the need to impress might have subverted the need to inform.

Although recognizing this versatility for dual roles should enhance the informed map viewer's healthy skepticism about the map author's expertise or motives, neither this recognition nor the map's demonstrated ability to distort and mislead should detract from an appreciation of the map's power to explore and explain geographic facts. White lies are an essential element of cartographic language, an abstraction with enormous benefits for analysis and communication. Like verbal language and mathematics, though, cartographic abstraction has costs as well as benefits. If not harnessed by knowledge and honest intent, the power of maps can get out of control.

LATITUDE AND LONGITUDE

Flattened at the poles and bulging slightly at the equator, the earth is not a perfect sphere. (As on the outside of a carousel, centrifugal force is strongest along the equator, where the radius of rotation is greatest. The rotating planet deforms like a highly viscous fluid—like a ball of clay on a potter's wheel.) Geodesists have found that the radius from the center of the earth to either pole is about 1/300 shorter than the radius to the equator. Although large-scale maps must take this deformation into account, small-scale maps of states, countries, or the entire world can safely and conveniently treat the earth as a sphere.

Latitude and longitude describe positions on the sphere. Longitude is somewhat arbitrary, anchored by international agreement, whereas latitude is a natural coordinate, related to the earth's rotation about its axis. Figure A.1 shows the equator in a plane through the center of the earth and perpendicular to the axis. Point A lies in another plane, parallel to the equator and also perpendicular to the axis; the circle formed where this plane intersects the sphere is a *parallel*. Like all points on the sphere, point A lies in one and only one parallel. *Latitude* is the angle, measured north or south from the equator, that identifies a particular parallel. It ranges from 0° at the equator to 90° at the poles and requires the letter N or S to establish its position north or south of the equator. Chicago, for example, is at 42° N, whereas Sydney, Australia, is at 34° S.

The equator is a *great circle*, the largest circle that occurs on a spherical surface. A great circle divides the sphere into two equal parts and describes the shortest-distance route between any two points along its circumference. A sphere has an infinite number of great circles, but only the equator is equidistant

North Pole

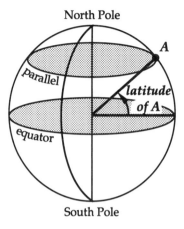

South Pole

FIGURE A.1. Spherical earth showing equator, poles, and a parallel and its latitude.

North Pole

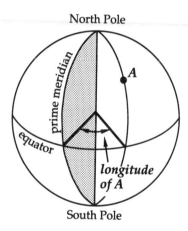

South Pole

FIGURE A.2. Spherical earth showing a meridian, its longitude, and the prime meridian.

from the poles. Except for the equator, the parallels are *small circles,* defined simply as any circle smaller than the equator. *Meridians,* which intersect the parallels at right angles, are halves of great circles running from pole to pole, as figure A.2 shows. Except at the poles, each point on the sphere has a single meridian and a single parallel.

Longitude, the other spherical coordinate, is the angle that identifies a meridian. Measured east or west as indicated by the letter E or W, it ranges from 0° at the *prime meridian* to 180° at the approximate location of the international date line. New York City, for example, is at 74° W, whereas the longitude of Moscow is about 38° E. In 1884, at the International Meridian Conference held in Washington, D.C., twenty-two of the twenty-five nations represented endorsed a prime meridian through the Royal Observatory at Greenwich, England. Subsequent international acceptance of the Greenwich meridian ended an era of cartographic isolation marked by prime meridians through Cadiz, Christiania, Copenhagen, Ferro, Lisbon, Naples, Paris, Pulkowa, Rio de Janeiro, and Stockholm, among others. Historians and other users of old maps must be particularly conscious of prime meridians and longitude.

SELECTED READINGS FOR FURTHER EXPLORATION

Readers interested in further information about maps and their uses might find this varied list of sources helpful.

Elements of the Map

Bunge, William. *Theoretical Geography.* Lund Studies in Geography, series C, General and Mathematical Geography, no. 1. Lund, Sweden: C. W. K. Gleerup, 1962. Especially chapter 2, "Metacartography."

Dent, Borden D. *Cartography: Thematic Map Design.* 3d ed. Dubuque, Iowa: William C. Brown, 1993.

Greenhood, David. *Mapping.* Chicago: University of Chicago Press, 1964.

Keates, J. S. *Understanding Maps.* New York: John Wiley, 1982.

MacEachren, Alan M. *Some Truth with Maps: A Primer on Symbolization and Design.* Washington, D.C.: Association of American Geographers, 1994.

———. *How Maps Work: Representation, Visualization, and Design.* New York: Guilford Press, 1995.

Monmonier, Mark. *Mapping It Out: Expository Cartography for the Humanities and Social Sciences.* Chicago: University of Chicago Press, 1993.

Robinson, Arthur H., and Barbara Bartz Petchenik. *The Nature of Maps: Essays toward Understanding Maps and Mapping.* Chicago: University of Chicago Press, 1976.

Robinson, Arthur H., Joel L. Morrison, Phillip C. Muehrcke, A. Jon Kimerling, and Stephen C. Guptill. *Elements of Cartography.* 6th ed. New York: John Wiley, 1995.

Snyder, John P. *Map Projections—A Working Manual.* U.S. Geological Survey Professional Paper 1395. Washington, D.C.: U. S. Government Printing Office, 1987.

History of Cartography

Blakemore, M. J., and J. B. Harley. *Concepts in the History of Cartography: A Review and Perspective.* Cartographica Monographs, no. 26. Toronto: University of Toronto Press, 1980.

Harley, J. B., and David Woodward, eds. *The History of Cartography*. 6 vols. Chicago: University of Chicago Press, 1987– .

Monmonier, Mark. *Maps with the News: The Development of American Journalistic Cartography*. Chicago: University of Chicago Press, 1989.

Robinson, Arthur H. *Early Thematic Mapping in the History of Cartography*. Chicago: University of Chicago Press, 1982.

Snyder, John P. *Flattening the Earth: Two Thousand Years of Map Projections*. Chicago: University of Chicago Press, 1993.

Thrower, Norman J. W. *Maps and Civilization: Cartography in Culture and Society*. 2d ed. Chicago: University of Chicago Press, 1995.

Woodward, David, ed. *Art and Cartography: Six Historical Essays*. Chicago: University of Chicago Press, 1987.

Map Use and Map Appreciation

Kjellstrom, Bjorn. *Be Expert with Map and Compass: The Orienteering Handbook*. New York: Charles Scribner's Sons, 1976.

Makower, Joel, ed. *The Map Catalog: Every Kind of Map and Chart on Earth and Even Some above It*. Rev. ed. New York: Vintage Books, 1992.

Maling, D. H. *Measurements from Maps: Principles and Methods of Cartometry*. Oxford: Pergamon Press, 1989.

Monmonier, Mark, and George A. Schnell. *Map Appreciation*. Englewood Cliffs, N.J.: Prentice-Hall, 1988.

Muehrcke, Phillip C., and Juliana O. Muehrcke. *Map Use: Reading, Analysis, and Interpretation*. 3d ed. Madison, Wis.: JP Publications, 1992.

Map Generalization

Eckert, Max. "On the Nature of Maps and Map Logic." Translated by W. Joerg. *Bulletin of the American Geographical Society* 40 (1908): 344–51.

Jenks, George F. "Lines, Computers, and Human Frailties." *Annals of the Association of American Geographers* 71 (1981): 1–10.

McMaster, Robert B., and K. Stuart Shea. *Generalization in Digital Cartography*. Washington, D.C.: Association of American Geographers, 1992.

Muller, Jean-Claude. "Generalization of Spatial Data Bases." In *Geographical Information Systems: Principles and Applications*, edited by David Maguire, Michael Goodchild, and David Rhind, 457–75. London: Longman, 1991.

Wright, John K. "Map Makers Are Human: Comments on the Subjective in Maps." *Geographical Review* 32 (1942): 527–44.

Development Maps and Geographic Information Systems

Aberley, Doug, ed. *Boundaries of Home: Mapping for Local Empowerment*. Philadelphia, Pa.: New Society Publishers, 1993.

Armstrong, Marc P., and others. "Cartographic Displays to Support

Locational Decision Making," *Cartography and Geographic Information Systems* 19 (1992): 154–64.

Goodchild, Michael F., Bradley O. Parks, and Louis T. Steyaert. *Environmental Modeling with GIS*. New York: Oxford University Press, 1993.

Pickles, John, ed. *Ground Truth: The Social Implications of Geographic Information Systems*. New York: Guilford Press, 1995.

Maps for Advertising and Political Propaganda

Bassett, Thomas J. "Cartography and Empire Building in Nineteenth-Century West Africa." *Geographical Review* 84 (1994): 316–35.

Boggs, S. W. "Cartohypnosis." *Scientific Monthly* 64 (1947): 469–76.

Davis, Bruce. "Maps on Postage Stamps as Propaganda." *Cartographic Journal* 22 (1985): 125–30.

Fleming, Douglas K. "Cartographic Strategies for Airline Advertising." *Geographical Review* 74 (1984): 76–93.

Gilmartin, Patricia. "The Design of Journalistic Maps: Purposes, Parameters, and Prospects." *Cartographica* 22 (Winter 1985): 1–18.

McDermott, Paul D. "Cartography in Advertising." *Canadian Cartographer* 6 (1969): 149–55.

Monmonier, Mark. "The Rise of the National Atlas." *Cartographica* 31 (Spring 1994): 1–15.

———. *Drawing the Line: Tales of Maps and Cartocontroversy*. New York: Henry Holt and Co., 1995.

Pickles, John. "Texts, Hermeneutics, and Propaganda Maps." In *Writing Worlds: Text, Metaphor, and Discourse*, edited by Trevor J. Barnes and James S. Duncan, 193–230. London: Routledge, 1992.

Quam, Louis O. "The Use of Maps in Propaganda." *Journal of Geography* 42 (1943): 21–32.

Speier, Hans. "Magic Geography." *Social Research* 8 (1941): 310–30.

Tyner, Judith A. "Persuasive Cartography." *Journal of Geography* 81 (1982): 140–44.

Maps, Defense, and Disinformation

Clarke, Keith C. "Maps and Mapping Technologies of the Persian Gulf War." *Cartography and Geographic Information Systems* 19 (1992): 80–87.

Demko, G. J., and W. Hezlep. "USSR: Mapping the Blank Spots." *Focus* 39 (Spring 1989): 20–21.

Stommel, Henry. *Lost Islands: The Story of Islands That Have Vanished from Nautical Charts*. Vancouver: University of British Columbia Press, 1984. Especially chapter 4, "The Fake Island of Captain Benjamin Morrell."

National Mapping and the Cartographic Subculture

Edney, Matthew H. "Politics, Science, and Government Mapping Policy in the United States, 1800–1925." *American Cartographer* 13 (1986): 295–306.

———. "Cartography without 'Progress': Reinterpreting the Nature and Historical Development of Mapmaking." *Cartographica* 30 (Summer/Autumn 1993): 54–68.

Harley, J. B. *Maps and the Columbian Encounter.* Milwaukee, Wis.: Golda Meir Library, University of Wisconsin, 1990.

Larsgaard, Mary Lynette. *Topographic Mapping of the Americas, Australia, and New Zealand.* Littleton, Colo.: Libraries Unlimited, 1984.

Lewis, G. Malcolm, "Metrics, Geometries, Signs, and Language: Sources of Cartographic Miscommunication between Native and Euro-American Cultures in North America." *Cartographica* 30 (Spring 1993): 98–106.

McHaffie, P. H. "The Public Cartographic Labor Process in the United States: Rationalization Then and Now." *Cartographica* 30 (Spring 1993): 55–60.

Rundstrom, Robert A. "The Role of Ethics, Mapping, and the Meaning of Place in Relations between Indians and Whites in the United States." *Cartographica* 30 (Spring 1993): 21–28.

Southard, R. B. "The Development of U.S. National Mapping Policy." *American Cartographer* 10 (1983): 5–15.

Thompson, Morris M. *Maps for America: Cartographic Products of the U.S. Geological Survey and Others.* Reston, Va.: U.S. Geological Survey, 1979.

Woodward, David. "Map Design and the National Consciousness: Typography and the Look of Topographic Maps," *Technical Papers of the American Congress on Surveying and Mapping* (Spring 1992): 339–347.

Data Maps

Fisher, Howard T. *Mapping Information: The Graphic Display of Quantitative Information.* Cambridge, Mass.: Abt Books, 1982.

Monmonier, Mark. "The Hopeless Pursuit of Purification in Cartographic Communication: A Comparison of Graphic-arts and Perceptual Distortions of Graytone Symbols." *Cartographica* 17 (1980): 24–39.

Tobler, W. R. "Choropleth Maps without Class Intervals?" *Geographical Analysis* 5 (1973): 262–65.

Tufte, Edward R. *The Visual Display of Quantitative Information.* Cheshire, Conn.: Graphics Press, 1983.

———. *Envisioning Information.* Cheshire, Conn.: Graphics Press, 1990.

Color

Brewer, Cynthia A. "Color Chart Use in Map Design." *Cartographic Perspectives,* no. 4 (1989): 3–10.

————. "Color Use Guidelines for Mapping and Visualization." In *Visualization in Modern Cartography*, edited by Alan M. MacEachren and D. R. Fraser Taylor, 123–47. Oxford: Pergamon, 1994.

English-Zemke, Patricia. "Using Color in Online Marketing Tools." *IEEE Transactions on Professional Communication* 31 (1988): 70–74.

Falk, David S., Dieter R. Brill, and David G. Stork. *Seeing the Light: Optics in Nature, Photography, Color, Vision, and Holography.* New York: Harper and Row, 1986. Especially pp. 238–86.

Hardin, C. L. *Color for Philosophers: Unweaving the Rainbow.* Indianapolis, Ind.: Hackett Publishing Co., 1988.

Hoadley, Ellen D. "Investigating the Effects of Color." *Communications of the ACM* 33 (1990): 120–25, 139.

Olson, Judy M. "Color and the Computer in Cartography." In *Color and the Computer*, edited by H. John Durrett, 205–19. Boston: Academic Press, 1987.

Multimedia and Electronic Cartography

Hall, Stephen S. *Mapping the Next Millennium.* New York: Random House, 1992.

Monmonier, Mark. *Technological Transition in Cartography.* Madison, Wis.: University of Wisconsin Press, 1985.

————. "Authoring Graphic Scripts: Experiences and Principles." *Cartography and Geographic Information Systems* 19 (1992): 247–60.

Peterson, Michael P. *Interactive and Animated Cartography.* Englewood Cliffs, N.J.: Prentice-Hall, 1995.

Sources of Illustrations

3.4 (left): U.S. Geological Survey, 1973, Northumberland, Pa., 7.5-minute quadrangle map.

3.4 (right): U.S. Geological Survey, 1969, Harrisburg, Pa., 1:250,000 series topographic map.

3.5 (left): New York State Department of Transportation, 1980, Rochester East (north part), 1:9,600 urban area map.

3.7: Metrorail System Map, courtesy of Washington Metropolitan Area Transit Authority.

4.1: Compiled from maps in Brian J. Hudson, "Putting Granada on the Map," *Area* 17, no. 3 (1985): 233–35.

7.1: Christopher Saxton, *Atlas of England and Wales* (1579), and Maurice Bouguereau, *Le théâtre françoys* (1594), atlases in the collection of the U.S. Library of Congress.

7.2: *Area Handbook for Pakistan* (Washington, D.C.: U.S. Government Printing Office, 1965), p. xii.

7.3 (upper): *Kashmir* (brochure) (New Delhi: Department of Tourism, Government of India, 1981), back cover.

7.3 (lower): *Pakistan Hotel Guide* (Karachi: Tourism Division, Government of Pakistan, 1984), back cover.

7.8 Courtesy of John P. Snyder (according to Snyder, the projection was devised about 1945 and first shown to cartographers in 1987; Snyder has not published it).

7.9: *Facts in Review* 1, no. 17 (8 December 1939): 1.

7.10: *Facts in Review* 3, no. 16 (5 May 1941): 250.

7.11: *Facts in Review* 2, no. 5 (5 February 1940): 33.

7.12: *Facts in Review* 2, no. 46 (30 November 1940): 566.

7.13: *Facts in Review* 2, no. 45 (25 November 1940): 557.

7.14: *Facts in Review* 2, no. 28 (8 July 1940): 294.

7.15: *Facts in Review* 1, no. 16 (30 November 1939): 3.

7.16: *Facts in Review* 3, no. 13 (10 April 1941): 182.

7.18: Associated Press Wirephoto map, used on the front page of the *Syracuse Herald-Journal* for 6 July 1950; reprinted with the permission of the Associated Press.

7.20: David M. Smith, *Where the Grass Is Greener* (Baltimore, Md.: Johns Hopkins University Press, 1982), fig. 1.6, © David M. Smith, 1979; reprinted by permission.

8.1: Compiled from facsimile maps in "Soviet Cartographic Falsification," *Military Engineer* 62, no. 410 (November-December 1970): 389–91, figs. 2a–e.

8.2: Compiled from facsimile maps in "Soviet Cartographic Falsification," *Military Engineer* 62, no. 410 (November-December 1970): 389–91, figs. 1a–c.

8.3: New York State Department of Transportation, 1978, Rome (north), 1:9,600 urban area map.

8.4 (map): U.S. Geological Survey, 1985, Blue Ridge Summit, Md.–Pa., 7.5-minute quadrangle map.

8.4 (quotation): Edward C. Papenfuse et al., *Maryland: A New Guide to the Old Line State* (Baltimore: Johns Hopkins University Press, 1976), p. 64.

8.5 (left): U.S. Army Map Service, 1946, Tonawanda West, N.Y., 7.5-minute quadrangle map.

8.5 (right): U.S. Geological Survey, 1980, Tonawanda West, N.Y., 7.5-minute quadrangle map.

9.1: U.S. Geological Survey, 1902, Everett, Pa., 15-minute quadrangle map.

9.4 (left): U.S. Geological Survey, 1980, Bunker Hill, Ind., 7.5-minute quadrangle map (photorevised).

9.4 (right): U.S. Geological Survey, 1994, Bunker Hill, Ind., 7.5-minute quadrangle map.

9.5: U.S. Geological Survey, 1982, North Haven East, Me., 7.5-minute quadrangle map (provisional).

9.6: U.S. Geological Survey, 1986, Scranton, Pa., 1:100,000 planimetric map.

9.7: U.S. Geological Survey, 1950, 1957, 1965, and 1969 editions, Harrisburg, Pa., 1:250,000 topographic map.

9.8 (above): U.S. Geological Survey, 1958, Cajon Mesa, Utah–Colo., 15-minute quadrangle map.

INDEX